BUFFALO SOLDIERS

CLASSICS *in* BLACK STUDIES EDITORIAL BOARD

CLASSICS IN BLACK STUDIES

BUFFALO SOLDIERS

THE COLORED REGULARS IN THE UNITED STATES ARMY

T. G. Steward

FOREWORD BY

Frank N. Schubert

HB

Humanity Books

an imprint of Prometheus Books
59 John Glenn Drive, Amherst, New York 14228-2197

Cover image: Courtesy of Library of Congress
Prints and Photographs Division, Gladstone Collection, LC-USZC4-6161

Published by Humanity Books, an imprint of Prometheus Books

Inquiries should be addressed to
Humanity Books
59 John Glenn Drive
Amherst, New York 14228–2197
VOICE: 716–691–0133, ext. 207
FAX: 716–564–2711
WWW.PROMETHEUSBOOKS.COM

07 06 05 04 5 4 3 2

Library of Congress Cataloging-in-Publication Data

Steward, T. G. (Theophilus Gould), 1843–1924.
[Colored regulars in the United States Army]
Buffalo soldiers : the colored regulars in the United States Army / by T.G. Steward ; foreword by Frank N. Schubert.
p. cm. — (Classics in Black studies)
Originally published: The colored regulars in the United States Army. Philadelphia : A.M.E. Book Concern, 1904.
Includes bibliographical references.
ISBN 1–59102–107–3 (pbk. : alk. paper)
1. Spanish-American War, 1898—Participation, African American. 2. African American soldiers—Cuba—History—19th century. 3. African American soldiers—History. I. Title. II. Series.

E725.5.N3S8 2003
973.8'933—dc21 2003051089

Printed in the United States of America on acid-free paper

THEOPHILUS GOULD STEWARD was born in Gouldtown, New Jersey, on April 17, 1843. As the son of free Blacks reared in a family that stressed education, he received his formal education in the Gouldtown public schools. He was ordained a minister in the African Methodist Episcopal Church in 1863. Following the Civil War, Steward helped organize the A.M.E. Church in South Carolina and Georgia. He was also active in Reconstruction politics in Georgia. After the war he graduated from the Episcopal Divinity School of Philadelphia, and later was awarded a doctor of divinity degree from Wilberforce University in Wilberforce, Ohio, in 1881.

From 1872 to 1891 Steward established a church in Haiti and preached in the eastern United States. In 1891 he joined the Twenty-fifth U.S. Colored Infantry, serving as its chaplain until 1907. Between 1907 and his death on January 11, 1924, Steward was a professor of history, French, and logic at Wilberforce University.

Works by Steward include: *Genesis Re-read* (1885); *Active Service, or Religious Work Among U.S. Soldiers* (1897); *A Charleston Love Story* (1899); *The Colored Regulars in the United States Army* (1904); *The Haitian Revolution 1791 to 1804, or Sidelights on the French Revolution* (1914); and *Fifty Years in the Gospel Ministry* (1920).

FOREWORD

The Civil War initiated many important changes in American society. Some, like the end of white supremacy in the South, were a very long time coming. Others, like the abolition of chattel slavery, were immediate and profound. Still others, such as the incorporation of black soldiers into the regular army, were less spectacular but still had long-term significance for the evolving place of African Americans in a largely white society.

Nearly 10 percent of the two million men who served in the army that defeated secession were black. Their contribution to the preservation of the Union helped pave the way for inclusion of black soldiers in the postwar regular army. The regiments that grew out of the army reorganizations of 1866 and 1869, the Ninth and Tenth Cavalry and the Twenty-fourth and Twenty-fifth Infantry, performed well on the frontier, became widely known as "Buffalo Soldiers" in the twentieth century, and in the two generations after the civil rights revolution of the 1960s gradually took on a mythical identity which held that they were the most effective and most competent units in the army of their day.[1]

These four regiments, while making their mark as hard-fighting frontier troops, also provided a home to the first few black line officers allowed to achieve commissioned rank in the regular force. The trickle started in 1879, when Henry O. Flipper survived four years of brutal treatment at the United States Military Academy at West Point to be commissioned a second lieutenant in the Tenth Cavalry. In the next nineteen years, only two other line officers, John Alexander and Charles Young, followed Flipper.

By the beginning of the conflict with Spain in 1898, five chaplains had also received commissions as captains, starting with Henry V. Plummer of the Ninth Cavalry in 1884 and Allen Allensworth of the Twenty-fourth Infantry in 1886. Both were Civil War veterans. They were also the first black ministers to wear the shepherd's crook insignia on the shoulder straps of their tunics as regular chaplains, but they followed numerous black clergymen who had served with African American units during the Civil War on a wartime volunteer basis.

As chaplains of black regiments, Plummer and Allensworth had responsibilities that went beyond attending to the spiritual needs of their men to include providing their common school education. When Congress authorized black regiments in the late 1860s, the legislation provided that their chaplains would also conduct schools for the soldiers, taking into account that many of the men had been slaves and lacked schooling. The law also assigned the chaplains of black regiments to the regimental headquarters and required that they remain with their regiments. This practice contrasted with that of assigning white chaplains to specific posts rather than to specific regiments.

In 1891 Theophilus Gould Steward became the third African American minister to receive an appointment as chaplain. He was forty-eight years old, an experienced minister and creative thinker who had already made an impact on the African Methodist Episcopal (A.M.E.) Church as well as on the post–Civil War South. Unlike his predecessors, who were both Baptist preachers and had been born into bondage in slave states, Steward came from a long-established community of free blacks, Gouldtown, in the southern tip of New Jersey. His father, James Steward, had immigrated to the United States from Haiti as a child in 1824 and soon afterward lost his parents. James Steward was reared in New Jersey by the uncle of his future wife, Rebecca Gould. She was descended from the founder of Gouldtown, black Benjamin Gould, who in 1683 had married Elizabeth Adams, the white granddaughter of English Lord John Fenwick. Chaplain Steward was the fourth of six children in a family that emphasized learning and independent thought as well as religion.[2]

Only eighteen when he became a minister, Theophilus Steward was in his early twenties when he went to Charleston, South Carolina, in 1865 and became involved in black community life in the former Confederacy. He accompanied Bishop Daniel Payne, a prominent A.M.E. figure who had already gained renown at Wilberforce University as the first black president of a black institution of higher learning. Slave states had encouraged black attendance at white churches, but churches run by blacks had been illegal in antebellum South Carolina. In the months just after the Civil War, Steward worked with Payne to establish the South Carolina Conference of their church, which became the center for A.M.E. expansion throughout the region and a major element of the church's effort to minister to the freedmen.[3]

Steward stayed in the South until 1872. After establishing and pastoring a church in Beaufort, South Carolina, he moved to Georgia, where he organized congregations and schools in small towns before serving a congregation in Macon. He also became active in state politics, helping to write the Republican Party's state platform in 1868 and leading a successful black protest against compulsory labor contracts in Americus, Georgia. While in Macon, Steward also worked as cashier of the black-owned Freedman's bank. In addition, he met and married Elizabeth Gadsden, with whom he had eight children.

In the next period of his professional life, from 1872 to 1891, Steward served several larger churches, established himself as a scholar, and began a prolific writing career. After a missionary trip to Haiti in 1873, during which he established a church in Port-au-Prince, he served churches in Brooklyn, Philadelphia, Wilmington (Delaware), and Baltimore, as well as Washington, D.C.'s prestigious Metropolitan A.M.E. Church in 1886–1887. He also graduated from Wilberforce in 1881 with the degree of doctor of divinity, and nearly became the university's president three years later, despite having made numerous enemies. Steward had criticized his church's liturgy as excessively formalistic and challenged its orthodoxy with some of his writings. He also had opposed the nomination to bishop of the famous advocate of the return of black Americans to Africa, Henry McNeal Turner, because of allegations of adultery. The combined effects of these actions cost him the position.[4]

Among Theophilus Steward's many publications, *Genesis Re-read*, which appeared in 1885, established his reputation in theology. This work, grounded in the theory of evolutionary theism, placed the evolutionary process within the context of a divine plan. As Alan Lamm summarized, Steward held that "evolution was God's way of doing business. . . ." This work put him among liberal thinkers trying to reconcile Christianity with the new scientific thought. It also set him apart from most Methodist ministers of the time and eventually enhanced his standing, especially when *Genesis Re-read* became a textbook at Payne Theological Seminary, a branch of Wilberforce University.[5]

Steward became chaplain of the Twenty-fifth Infantry in 1891, and served the regiment for sixteen years, until 1907. Appointments to chaplaincies, prized by black ministers for the officer's pay and stability that came with them, depended on political influence, for white as well as for black clergymen. Steward's main supporter was Philadelphia department-store owner John Wanamaker, who was President Benjamin Harrison's Postmaster General. Steward also had the backing of former senator Blanche K. Bruce of Mississippi and the celebrated abolitionist, publisher, and orator Frederick Douglass.[6]

Steward was separated from Plummer and Allensworth by more than the circumstances and locations of their birth. The three differed widely in styles and interests. Allensworth emphasized education, generally on a practical vocational level that reflected his support for the philosophy of Booker T. Washington, and sought ways to enlist better recruits for the army. Plummer was a dynamic, charismatic preacher who favored mass revival meetings and temperance lectures and objected to services of competing denominations while serving the Ninth at Fort Robinson, Nebraska. He also corresponded with Bishop Turner and promoted African colonization schemes. Steward tended to stress close personal contact and dialogue with individual soldiers and established close working relationships with civilian ministers in nearby towns, regardless of denomination.[7] Although he was known by 1891 as "one of the most scholarly ministers of the A.M.E. Church," he was never aloof from the soldiers of his regiment.[8] In fact, he seemed to two modern scholars "to have had a closer rapport with enlisted men than any of the other chaplains, black or white."[9]

After Steward joined the Twenty-fifth Infantry at Fort Missoula, Montana, he concentrated on earning the respect of the soldiers and officers with whom he served. While struggling to come to terms with the death of his wife in 1893, he learned to ride horses and mules, went fishing and hunting, and took his turn tending the post vegetable gardens, probably the only chaplain in the service who so involved himself in military routine. As Steward himself wrote, he "became possessed of the army spirit and identified myself with its discipline and training as well as its outdoor life." His efforts did not go unappreciated. One commander, Col. George L. Andrews, called him "well educated, gentlemanly, refined and respected by all." Another, Col. Andrew S. Burt, called him "the most conscientious chaplain in the discharge of his duties I have ever served with," and "an ornament to the service."[10]

In the 1890s, while becoming immersed in the active life of a regimental chaplain on the northern plains, Steward's published works focused increasingly on his new military vocation. His book *Active Service: or Religious Work Among US Soldiers* appeared in 1897. It contained sixteen articles on aspects of the ministry among soldiers, twelve of them by chaplains and another by Maj. Gen. Oliver Otis Howard, head of the Freedmen's Bureau during Reconstruction. Steward edited the volume and wrote the preface. His articles about military life and issues, covering such matters as confrontations between soldiers and striking civilian workers and whether alcohol-serving canteens had a place on army posts, appeared in general circulation and professional military magazines.

Before the decade ended, in 1896 Steward remarried. His second wife, Susan Maria McKinney Steward, was one of the first black women in the United States to complete medical school. For many years she was the resident physician at Wilberforce. She was also an author and musician.

In the spring of 1898, after the United States declared war on Spain, the Twenty-fifth Infantry was among the first units to make the long train trip from western posts to southeastern assembly areas. Soon much of the army headed in the same direction to camps near the main port of debarkation in Tampa, Florida. Steward went with his regiment as far as Chattanooga and nearby Chickamauga Military Park, where many regiments camped and awaited further orders. Like the other black chaplains—George Prioleau of the Ninth Cavalry, William T. Anderson of the Tenth Cavalry, and Allensworth of the Twenty-fourth Infantry—Steward remained in the United States while his regiment went to war. On recruiting duty in Dayton, Ohio, near the Wilberforce community where he was well known, Steward sought black troops for the army while the soldiers of the Twenty-fifth boarded transports in Tampa and sailed for Cuba.

All four black regiments fought in Cuba. During the main battle of July 1, just outside of Santiago, the Ninth, Tenth, and Twenty-fourth participated in the attack on San Juan Heights, and the Twenty-fifth was among the units in the assault on El Caney, just to the north. The newspaper reports from the front featured the exploits of two distinct groups, the four regiments of black regulars and the First U.S. Volunteer Cavalry, popularly known as the Rough Riders and commanded during the fighting by Lt. Col. Theodore Roosevelt. The country was abuzz with the exploits of both. All of a sudden, while the status of blacks was deteriorating severely, with segregation becoming the law of much of the land and racial violence becoming almost commonplace, the United States had a group of widely recognized black heroes.[11]

Chaplain Steward rejoined his regiment in September 1898 at the rest camp near Montauk Point on the eastern end of Long Island, New York. He may already have been thinking about writing a book about the exploits of the African American soldiers in Cuba. In December, just nine days after the Treaty of Paris ended the war with Spain and left the remnants of Spain's once vast empire—Cuba, Puerto Rico, Guam, and the Philippines—in the hands of the United States, Steward applied to the Adjutant General to be placed on detached service to write *The Colored Regulars*. He wanted to tell future generations of black Americans the history of "the valorous conduct of the four black regiments in Cuba," and further justify the policy that kept blacks in the army, "for the cause of my race and my country." The army allowed him several months away from his regular duties, much of which he spent at Wilberforce. By the end of 1899 he was finished.[12] The book was published by the A.M.E. Church's Book Concern in 1904.[13]

Chaplain Steward declared his purpose in writing *The Colored Regulars in the United States Army* in the preface. "The work has been performed," he wrote, "with an earnest desire to obtain and present the truth, hoping that the reader will be inspired by it to a more profound respect for the brave and skilled black men who also passed through that severe baptism of fire and suffering, contributing their full share to their country's honor."[14] Steward believed that it was important to tell this story because military service was critical to the black quest for equality and respect. Carrying out the duties of a soldier ultimately gave a man a claim on the rights and privileges of citizenship. That was the bottom line: ". . . the actual discharge of the duties of the soldier in defense of the nation entitles one to all common rights, to the nation's gratitude, and to the highest honors for which he is qualified."[15] This was the same insight articulated by Frederick Douglass at the start of the Civil War, when he said, "Once let the black man get upon his person the brass letters 'US,' let him get an eagle on his button and a musket on his shoulder and bullets in his pockets and there is no power on earth which can deny that he has earned the right of citizenship in the United States." They were right, although only in the second half of the twentieth century did their prophecy become reality.

Steward saw the military history of black Americans as part of a broader theme in their history, one of militancy and resistance to wrong, even by force. These elements of the black past showed "a toughness of fibre and steadiness of purpose sufficient to make the backbone of a real history." He viewed the tradition of participation in the armed forces, from the American Revolution to the fight against secession and slavery, both of which were embodied in the Confederacy, as part of the same overall tradition of black struggle. Inside Nat Turner, the leader of a slave insurrection in Virginia during 1831, Steward saw "the elements of a vigorous captain."[16] "The present work," he wrote, referring to *The Colored Regulars*, "deals with these elements of character as they are exhibited in the garb of a soldier. . . ."[17]

After reprising black military history from the beginning of the Republic to the service of black regulars on the frontier, Steward turned to the events of 1898. He told the entire story of black soldiers in the Cuban campaign, from the movement of troops to assembly points in Georgia and Florida through the peace treaty. Along the way, he covered the major battles—Las Guasimas, San Juan, and El Caney—and the service of black infantrymen as volunteer nurses in the yellow fever hospital at Siboney, Cuba. Steward told the story judiciously and fairly, crafting a balanced narrative that gave credit to white and black regulars and to Lieutenant Colonel Roosevelt and the Rough Riders. Along the way, he did what he could to demolish stereotypes, showing the black regulars as cool professional soldiers and citing cases of black sergeants taking the initiative and leading their men into battle when their white officers had fallen. A century after its publication, *The Colored Regulars* remains a model of careful narrative history that still is the basic account of the role of the black regulars in the war against Spain.

One of the more remarkable features of the book is the war diary of Sgt. Maj. Edward L. Baker of the Tenth Cavalry.[18] Baker was a superb soldier, among the best of his generation of black regulars, and a recipient of the Medal of Honor for the rescue of a wounded comrade under fire in Cuba.[19] Steward considered Baker and another Medal of Honor holder, William McBryar of the Twenty-fifth, the two best candidates for regular army commissions, called both "old soldiers, highly intelligent, and gallant,"[20] and reprinted the Baker diary unfortunately without comment regarding its provenance. Although the original manuscript has never been found, both the style and the content of the published diary are consistent with Baker's other writings and his career.

Chaplain Steward's history was well received. The book carried an introductory endorsement by Lt. Gen. Nelson A. Miles, the commanding general of the army in 1898. The black press gave it extensive publicity, drawing explicit connections between black military service and the claims of African Americans to a broader role in the affairs of the nation. The *New York Age*, for example, greeted publication with an editorial railing against prejudicial treatment of blacks in the military and claiming that Steward sought to arouse the American people to a proper sense of gratitude and open the door to promotion for blacks.[21]

By the time that the book appeared, Steward had already served a tour of duty with his regiment in the Philippines and returned to the trans-Mississippi West where his military career had begun. His soldiers were glad to have him among them. As Private Rienzi Lemus wrote from the Philippines to his hometown Richmond, Virginia, newspaper, the *Planet*, "The boys were glad to welcome him as his literary work is a rare treat and his presence is much enjoyed after an extended absence." They may also have been impressed with his willingness to confront white soldiers who denied him the respect that was due his rank and refused to salute him. Steward was not afraid to take them to task publicly and complain to their commanders. When *The Colored Regulars* appeared in 1904, he was serving with the Twenty-fifth at Fort Niobrara in the sand hills of northern Nebraska.

During the spring of 1906, when the army considered sending the Twenty-fifth to participate in maneuvers in Texas before stationing the regiment in that state, he joined other officers in strongly advising against it. There had been violent encounters between black regulars and white militia members during the Texas maneuvers of 1903, and they feared further trouble. The army backed down and cancelled orders for the 1906 maneuvers, but transferred the Twenty-fifth to Texas anyway.

The transfer set the stage for the Brownsville affray of August 1906. After a nighttime shooting near the edge of town nearest to Fort Brown in which one civilian was killed, townspeople blamed the men of the Twenty-fifth. With only conflicting circumstantial evidence and no confessions from the soldiers, Pres. Theodore Roosevelt concluded that all of the black soldiers were involved in a conspiracy of silence to shield the perpetrators among them. Roosevelt consequently dismissed without honor the entire battalion, 167 men. Steward, who visited soldiers incarcerated after Brownsville, defended their innocence, but neither he nor other supporters of the soldiers convinced the president to overturn his decision.[22] In April 1907, at Fort Reno, Oklahoma, he reached the retirement age of sixty-four and his career as chaplain of the Twenty-fifth Infantry ended.

After military service, Steward returned to Wilberforce. As vice president of the university from 1908, as well as chaplain and professor of history, French, and logic, he returned to writing and teaching. He led Wilberforce fund drives in several mid-Western cities, and with his second wife represented the A.M.E. Church at the First Universal Races Conference, held at the University of London in July 1911.

During his years at Wilberforce, Steward wrote his study of the Haitian revolution, *The Haitian Revolution 1791 to 1804, or Sidelights on the French Revolution*, a book that was published in 1914 to positive reviews by American scholars. Historian Albert Bushnell Hart wrote that the book "cannot fail to be serviceable both for the understanding of the Negro race and the relations of France with the West Indies."[23] In this book, Steward elaborated on ideas that he had expressed in *The Colored Regulars*, developing a theory of education for black Americans which put him somewhere between Booker T. Washington's advocacy of job-oriented industrial training and W. E. B. Du Bois's academic emphasis. In a letter to the *Indianapolis Freeman*

(August 19, 1905) he urged that all schools for blacks employ ex-soldiers to teach drill and military matters to all male students. "Soft men," he argued, "cannot carry on a hard fight."[24] Professor Steward's analysis of the Haitian revolution, which built on his views regarding the relationship between military service and citizenship, supported this position. In Haiti, only the reality of blacks as victorious soldiers forced whites to shed their disdain. There, as elsewhere when oppressed peoples struggled for liberty, "swords precede plowshares and . . . the spear goes before the pruning hook."[25]

Steward's views of education have been generally ignored in the continuing emphasis on the views of Du Bois and Washington. However, he added a rationale that was important to the development of military programs, which were instituted at some black high schools and colleges at the turn of the twentieth century. These programs filled what he considered to be the large gap in the main competing views of education, encouraging "the development of moral fiber and force" that characterized military service. "There is no greater civilizing agency for the Negro," he insisted, "whether we look upon the conservative or advancing side, than the army; and it is through this instrumentality, amid the strife and blood soon to engulph [*sic*] more or less the civilized world, that I look for the American Negro to emerge from his present lot."[26]

Chaplain Steward had a long and productive life. His last major book, a memoir entitled *Fifty Years in the Gospel Ministry*, appeared in 1920, and he died four years later. His fellow black regular chaplains were all remarkable men, eloquent in the pulpit and articulate in print, and William Anderson was a medical doctor as well as a minister. However, Steward stood head and shoulders above them as a theologian and scholar. He also showed an intense commitment to direct personal relationships with the enlisted men of his regiment. If Steward had a significant shortcoming it was his deep faith in the army. Even after Brownsville, he persisted in the view that the army provided an excellent place for African Americans to seek opportunity. As Alan Lamm noted, "His only major blindspot was his unwillingness to see the inherent racism that existed within the army itself."[27]

In the early 1960s, as an undergraduate history student at Howard University, I took the required courses in African American history and learned of the injustice visited on the Twenty-fifth after Brownsville. Resolving to learn more about this troubling affair, I made my first visit to the National Archives. During a routine reference interview, the distinguished archivist and historian Sara Dunlap Jackson introduced me to Theophilus G. Steward. Jackson was taken aback by my ignorance of this eminent clergyman and scholar. She insisted that it was essential to read his book *The Colored Regulars* before undertaking research on Brownsville or any other topic involving black soldiers in the years after the Civil War. She was right. It was a basic book then, and it remains an important resource today. I never regretted following her advice.

Frank N. Schubert
Chief of Joint Operational History (Retired)
Joint History Office
Office of the Chairman
Joint Chiefs of Staff
The Pentagon

Notes

1. On the Buffalo Soldiers and their frontier service, see William H. Leckie, *The Buffalo Soldiers: A Narrative of the Negro Cavalry in the West* (Norman: University of Oklahoma Press, 1967); Arlen L. Fowler, *The Black Infantry in the West 1869–1891* (Westport, Conn.: Greenwood, 1971); William A. Dobak and Thomas D. Phillips, *The Black Regulars 1866–1898* (Norman: University of Oklahoma Press, 2001). On the modern myth of the Buffalo Soldier, see Frank N. Schubert, "Buffalo Soldiers: Myths and Realities," *Army History* (spring 2001): 13–18, also available on line at http://www.captainbuffalo.com/myths.html.

2. Alan K. Lamm, *Five Black Preachers in Army Blue: The Buffalo Soldier Chaplains* (Lewiston, N.Y.: Edwin Mellen, 1998), pp. 193, 195.

3. Ibid., p. 196.

4. Ibid., p. 201.

5. Frank N. Schubert, "Theophilus Gould Steward," *Dictionary of American Negro Biography*, ed. Rayford W. Logan and Michael R. Winston (New York: W. W. Norton, 1982), p. 570; Lamm, *Five Black Preachers in Army Blue*, pp. 198–99.

6. Dobak and Phillips, *The Black Regulars 1866–1898*, p. 115; Lamm, *Five Black Preachers in Army Blue*, p. 202.

7. Frank N. Schubert, *Buffalo Soldiers, Braves and the Brass: The Story of Fort Robinson, Nebraska* (Shippensburg, Pa.: White Mane, 1993), pp. 128–34; Dobak and Phillips, *The Black Regulars 1866–1898*, p. 118; Earl F. Stover, *Up from Handymen: The United States Army Chaplaincy, 1865–1920* (Washington, D.C.: Office of the Chief of Chaplains, Department of the Army, 1977), pp. 89–90; Marvin E. Fletcher, *The Black Soldier and Officer in the United States Army 1891–1917* (Columbia: University of Missouri Press, 1974), p. 104.

8. *Cleveland Gazette*, August 8, 1891.

9. Dobak and Phillips, *The Black Regulars 1866–1898*, p. 122.

10. Steward quoted from William Seraile, *Voice of Dissent:*

Theophilus Gould Steward (1843–1924) and Black America (New York: Carlson, 1991), pp. 112–13; Andrews quoted from Lamm, *Five Black Preachers in Army Blue*, p. 205; Burt quoted from Schubert, *On the Trail of the Buffalo Soldier: Biographies of African Americans in the U.S. Army* (Wilmington, Del.: Scholarly Resources, 1995), p. 405.

11. At the beginning of the twenty-first century, two competing views of the fighting of July 1, 1898, still dominate much of the literature, emphasizing either the role of the black regulars or Roosevelt's volunteers. On this issue and the overall impact of the black regulars on the black community and the general population at home, see Frank N. Schubert, "Buffalo Soldiers at San Juan Hill," *Army History* (summer 1998): 36–38, also available on line at http://www.captainbuffalo.com/sanjuan.html.

12. For excerpts of some of the correspondence regarding Steward's project and the support that he garnered, see Schubert, *On the Trail of the Buffalo Soldier*, p. 405.

13. The book was first reprinted with a short introduction by William Loren Katz by Arno Press and the *New York Times* in 1969 as part of the series "The American Negro, His History and Literature."

14. Theophilus G. Steward, *The Colored Regulars in the United States Army* (Philadelphia: A.M.E. Book Concern, 1904), preface.

15. Ibid., p. 32.

16. Ibid., p. 68.

17. Ibid., p. 17.

18. Ibid., pp. 255–79.

19. For the career of Edward Baker, see Frank N. Schubert, *Black Valor: Buffalo Soldiers and the Medal of Honor, 1870–1898* (Wilmington, Del.: Scholarly Resources, 1997), pp. 145–61.

20. Ibid., p. 155.

21. *New York Age*, June 8, 1905.

22. Fletcher, *The Black Soldier and Officer in the United States Army*, p. 119; Stover, *Up from Handymen*, p. 167. On the events at Brownsville, see John D. Weaver, *The Brownsville Raid* (New York: W. W. Norton, 1970).

23. Quoted in the *Cleveland Gazette*, July 10, 1915.

24. *Indianapolis Freeman*, August 19, 1905.

25. Theophilus G. Steward, *The Haitian Revolution 1791 to 1804, or Sidelights on the French Revolution* (New York: Neale Publishing, 1914), p. vi.

26. Lamm, *Five Black Preachers in Army Blue*, p. 211.

27. Ibid., p. 226.

Chaplain T. G. Steward, D. D.

BUFFALO SOLDIERS

TABLE OF CONTENTS.

PREFACE.

The material out of which the story of the COLORED REGULARS has been constructed has been collected with great pains, and upon it has been expended a serious amount of labor and care. All the movements of the Cuban campaign, and particularly of the battles, have been carefully studied by the aid of official reports, and conversations and correspondence with those who participated in them. The work has been performed with an earnest desire to obtain and present the truth, hoping that the reader will be inspired by it to a more profound respect for the brave and skilled black men who passed through that severe baptism of fire and suffering, contributing their full share to their country's honor.

It is also becoming in this place to mention with gratitude the encouragement given by the War Department both in granting me the time in which to do the work, and also in supplying me with documents and furnishing other facilities. By this enlightened course on the part of the Department great aid has been given to historical science, and, incidentally, very important service rendered to the cause of freedom and humanity. A struggling people has been helped and further glory reflected upon the Government. The President, himself, has manifested a kindly interest in the work, and has wished that the story of the black soldiers should be told to the world. The interest of the Commanding General of the Army is shown in his letter.

Thus encouraged from official sources and receiving the

most hearty words of cheer from friends, of whom none has been more potent or more earnest than Bishop B. W. Arnett, D. D., of the African M. E. Church, I have, after five months of severe labor, about completed my task, so far as I find it in my power to complete it; and trusting that the majesty and interest of the story itself will atone for any defects in the style of the narration, the volume is now offered to a sympathetic public, affectionately dedicated to the men whose heroic services have furnished the theme for my pen.

T. G. STEWARD.

Wilberforce, Ohio, September, 1899.

LETTER FROM GENERAL MILES.

Headquarters of the Army, Washington,
August 5, 1899.

Rev. T. G. Steward, Chaplain 25th Infantry,
Wilberforce, Ohio.

Dear Sir:—Your letter of the 20th ultimo was duly received, but my time has been so much engrossed with official duties, requiring my presence part of the time out of the city, that it has not been practicable to comply with your request earlier; and even now I can only reply very briefly.

You will remember that my acquaintance with negro character commenced during the Civil War. The colored race then presented itself to me in the character of numerous contrabands of war, and as a people who, individually, yearned for the light and life of liberty. Ages of slavery had reduced them to the lowest ebb of manhood. From that degree of degradation I have been an interested spectator of the marvelously rapid evolution of the down-trodden race. From the commencement of this evolution to the present time I have been more or less in a position to closely observe their progress. At the close of the war I was in command of one of the very important military districts of the South, and my concern for the welfare of all the people of that district, not excluding the people of color, you will find evidenced in the measures taken by me, more especially in regard to educational matters, at that time. The first regiment which I commanded on entering the Regular Army of the United States at the close of the war was made up of colored troops. That regiment—the 40th Infantry—achieved a reputation for military conduct which forms a record that may be favorably compared with the best regiments in the service. Then, again, refer to my General Order No. 1, issued after the fall of Santiago, and you will see that recognition is not grudgingly given to the troops who heroically fought there, whether of American, of African, or of Latin descent. If so early in the second generation of the existence of the race in the glorious light of liberty it produces such orators as Douglas, such educators as Booker T. Washington, such divines as the Afro-American Bishops, what may we not expect of the race when it shall have experienced as many generations of growth and development as the Anglo-Saxons who now dominate the thought, the inventive genius, the military prowess, and the commercial enterprise of the world! Very truly yours,

NELSON A. MILES.

Lieutenant-General Nelson A. Miles.

Headquarters of the Army,

Siboney, Cuba, July 16, 1898.

General Field Orders No. 1.

The gratifying success of the American arms at Santiago de Cuba and some features of a professional character both important and instructive, are hereby announced to the army.

The declaration of war found our country with a small army scattered over a vast territory. The troops composing this army were speedily mobilized at Tampa, Fla. Before it was possible to properly equip a volunteer force, strong appeals for aid came from the navy, which had inclosed in the harbor of Santiago de Cuba an important part of the Spanish fleet. At that time the only efficient fighting force available was the United States Army, and in order to organize a command of sufficient strength, the cavalry had to be sent dismounted to Santiago de Cuba with the infantry and artillery.

The expedition thus formed was placed under command of Major-General Shafter. Notwithstanding the limited time to equip and organize an expedition of this character, there was never displayed a nobler spirit of patriotism and fortitude on the part of officers and men going forth to mantain the honor of their country. After encountering the vicissitudes of an ocean voyage, they were obliged to disembark on a foreign shore and immediately engage in an aggressive campaign. Under drenching storms, intense and prostrating heat, within a fever-afflicted district, with little comfort or rest, either by day or night, they pursued their purpose of finding and conquering the enemy. Many of them, trained in the severe experience of the great war, and in frequent campaigns on the Western plains, officers and men alike exhibited a great skill, fortitude, and tenacity, with results which have added a new chapter of glory to their country's history. Even when their own generals in several cases were temporarily disabled, the troops fought on with the same heroic spirit until success was finally achieved. In many instances the officers placed themselves in front of their commands, and under their direct and skillful leadership the trained troops of a brave army were driven from the thickets and jungles of an almost inaccessible country. In the open field the troops stormed intrenched infantry, and carried and captured fortified works with an unsurpassed daring and disregard of death. By gaining commanding ground they made the harbor of Santiago untenable for the Spanish fleet, and practically drove it out to a speedy destruction by the American Navy.

While enduring the hardships and privations of such campaign, the troops generously shared their scanty food with the 5,000 Cuban patriots in arms, and the suffering people who had fled from the besieged city. With the twenty-four regiments and four batteries, the flower of the United States Army, were also three volunteer regiments. These though unskilled in warfare, yet, inspired with the same spirit, contributed to the victory, suffered hardships, and made sacrifices with the rest. Where all did so well, it is impossible, by special mention, to do justice to those who bore conspicuous part. But of certain unusual features mention cannot be omitted, namely, the cavalry dismounted, fighting and storming works as infantry, and a regiment of colored troops, who, having shared equally in the heroism as well as the sacrifices, is now voluntarily engaged in nursing yellow-fever patients and burying the dead. The gallantry, patriotism and sacrifices of the American Army, as illustrated in this brief campaign, will be fully appreciated by a grateful country, and the heroic deeds of those who have fought and fallen in the cause of freedom will ever be cherished in sacred memory and be an inspiration to the living.

By command of Major-General Miles:

J. C. GILMORE,
Brigadier-General, United States Volunteers.

INTRODUCTORY.

To write the history of the Negro race within that part of the western world known as the United States of America would be a task to which one might devote a life time and still fail in its satisfactory accomplishment. The difficulties lying in the way of collecting and unifying the material are very great; and that of detecting the inner life of the people much greater. Facts and dates are to history what color and proportion are to the painting. Employed by genius, color and form combine in a language that speaks to the soul, giving pleasure and instruction to the beholder; so the facts and dates occurring along the pathway of a people, when gathered and arranged by labor and care, assume a voice and a power which they have not otherwise. As these facts express the thoughts and feelings, and the growth, of a people, they become the language in which that people writes its history, and the work of the historian is to read and interpret this history for the benefit of his fellow men.

Borrowing a second illustration from the work of the artist, it may be said, that as nature reveals her secrets only to him whose soul is in deepest sympathy with her moods and movements, so a people's history can be discovered only by one whose heart throbs in unison with those who have made the history. To write the history of any people successfully one must read it by the heart; and the best part of history, like the best part of the picture, must ever remain unexpressed. The artist sees more, and feels more than he is able to transfer to his canvas, however entrancing his presentation; and the historian sees and feels more than his brightest pages convey to his readers. Nothing less than a profound respect and love for humankind and a special attraction toward a particular people and age, can fit one to engage in so sublime a task as that of translating the history of a people into the language of common men.

The history of the American Negro differs very widely from that of any people whose life-story has been told; and when it shall come to be known and studied will open an entirely new view of experience. In it we shall be able to see what has never before been discovered in history; to wit: the absolute beginning of a people. Brought to these shores by the ship-load as freight, and sold as merchandise; entirely broken away from the tribes, races, or nations of their native land; recognized only as African slaves, and forbidden all movement looking toward organic life; deprived of even the right of family or of marriage, and corrupted in the most shameless manner by their powerful and licentious oppressors—it is from this heterogeneous protoplasm that the American Negro has been developed. The foundation from which he sprang had been laid by piecemeal as the slave ships made their annual deposits of cargoes brought from different points on the West Coast, and basely corrupted as is only too well known; yet out of it has grown, within less than three hundred years, an organic people. Grandfathers and great-grandfathers are among them; and personal acquaintance is exceedingly wide. In the face of slavery and against its teaching and its power, overcoming the seduction of the master class, and the coarse and brutal corruptions of the baser overseer class, the African slave persistently strove to clothe himself with the habiliments of civilization, and so prepared himself for social organization that as soon as the hindrances were removed, this vast people almost immediately set themselves in families; and for over thirty years they have been busily engaged hunting up the lost roots of their family trees. We know the pit whence the Afro-American race was dug, the rock whence he was hewn; he was born here on this soil, from a people who in the classic language of the Hebrew prophet, could be described as, No People.

That there has been a majestic evolution quietly but rapidly going on in this mass, growing as it was both by natural development and by accretion, is plainly evident. Heterogeneous as were the fragments, by the aid of a common language and a common lot, and cruel yet partially civilizing control, the whole people were forced into a common outward form, and to a remarkable extent, into the same ways of thinking. The affinities within were really aided by the repulsions without, and when finally freed from slavery, for an ignorant and inexperienced people, they presented an astonishing spectacle of unity. Socially, politically and religiously, their power to work together showed itself little less than marvellous. The Afro-American, developing from this slave base, now directs great organizations of a religious character, and in comprehensive sweep invites to his co-operation the inhabitants of the isles of the sea and of far-off Africa. He is joining with the primitive, strong, hopeful and expanding races of Southern Africa, and is evidently preparing for a day that has not yet come.

The progress made thus far by the people is somewhat like that made by the young man who hires himself to a farmer and takes his pay in farming stock and utensils. He is thus acquiring the means to stock a farm, and the skill and experience necessary to its successful management at the same time. His career will not appear important, however, until the day shall arrive when he will set up for himself. The time spent on the farm of another was passed in comparative obscurity; but without it the more conspicuous period could never have followed. So, now, the American colored people are making history, but it is not of that kind that gains the attention of writers. Having no political organizations, governments or armies they are not performing those deeds of splendor in statesmanship and war over which the pen of the historian usually delights to linger. The people, living, growing, reading, thinking, working, suffering, advancing and dying—these are all common-place occurrences, neither warming the heart of the observer, nor capable of brightening the page of the chronicler. This, however, is, with the insignificant exception of Liberia, all that is yet to be found in the brief history of the Afro-American race.

The period for him to set up for himself has not yet come, and he is still acquiring means and training within a realm controlled in all respects by a people who maintain toward him an attitude of absolute social exclusion. His is the history of a people marching from nowhere to somewhere, but with no well-defined Canaan before them and no Moses to lead. It is indeed, on their part, a walk by faith, for as yet the wisest among the race cannot tell even the direction of the journey. Before us lie surely three possible destinies, if not four; yet it is not clear toward which one of these we are marching. Are we destined to see the African element of America's population blend with the Euro-American element and be lost in a common people? Will the colored American leave this home in which as a race he has been born and reared to manhood, and find his stage of action somewhere else on God's earth? Will he remain here as a separate and subordinate people perpetuating the conditions of to-day only that they may become more humiliating and exasperating? Or is there to arise a war of races in which the blacks are to be exterminated? Who knows? Fortunately the historian is not called upon to perform the duties of prophet. His work is to tell what has been; and if others, building upon his presentation of facts can deduce what is to be, it is no small tribute to the correctness of his interpretations; for all events are parts of one vast system ever moving toward some great end. One remark only need be made. It is reasonable to presume that this new Afro-American will somehow and somewhere be given an opportunity to express that particular modification of material life which his spiritual nature will demand. Whether that expression will be made here or elsewhere; whether it will be higher or lower than what now surrounds us, are questions which we may well leave to the future.

No people can win and hold a place, either as a nation among other nations, or as an elementary component of a nation, merely by its own goodness or by the goodness of others. The struggle for national existence is a familiar one, and is always initiated by a display of physical force. Those who have the power seize territory and government, and those who CAN, keep possession and control. It is in some instances the backing up of right by might, and in others the substituting of right by might. Too often the greatest of all national crimes is to be weak. When the struggle is a quiet one, going on within a nation, and is that of an element seeking a place in the common social life of the country, much the same principles are involved. It is still a question to be settled by force, no matter how highly the claim of the weaker may be favored by reason and justice.

The powers by which a special people may emerge from an unhappy condition and secure improved social relations, using the word social in its broadest sense, are physical, intellectual and material. There must be developed manly strength and courage and a power of intellect which will manifest itself in organization and attractiveness, and in the aptitude of employing appropriate methods for ends in view. To these must be added the power that comes through wealth; and thus, with the real advancement of condition and character will come, tardily and grudgingly perhaps, but nevertheless surely, improved social standing. Once filled with the common national spirit, partaking of its thoughts, entering heartily into the common movements, having the same dress, language and manners as others, and being as able and as willing to help as to be helped, and withal being in fact the most intensely American element on the continent because constructed on this soil, we may hope that the Afro-American will ultimately win and hold his proper place.

The history made by the American Negro has been so filled with suffering that we have overlooked the active side. The world has heard so much of the horrors of the "Middle Passage"; the awful sufferings of the slave; the barbarous outrages that have been perpetrated upon ex-slaves; the inhuman and senseless prejudices that meet colored Americans almost everywhere on their native soil; that it has come to look upon this recital as the whole of the story. It needs to be told that these records constitute the dark side of the picture, dark and horrible enough, to be sure, but this is by no means the whole picture. If there are scenes whose representations would serve to ornament the infernal regions, pictures over which fiends might gloat, there are also others which angels might delight to gaze upon. There has been much of worthy action among the colored people of this country, wherever the bonds of oppression have been slackened enough to allow of free movement. There have been resistance to wrong by way of remonstrance and petition, sometimes even by force; laudable efforts toward self-education; benevolent and philanthropic movements; reform organizations, and commendable business enterprise both in individuals and associations. These show a toughness of fibre and steadiness of purpose sufficient to make the backbone of a real history.

The present work deals with these elements of character as they are exhibited in the garb of the soldier. When men are willing to fight and die for what they hold dear, they have become a moving force, capable of disturbing the currents of history and of making a channel for the stream of their own actions. The American Negro has evolved an active, aggressive element in the scientific fighting men he has produced. Individual pugilists of that race have entered all classes, from featherweight to heavyweight, and have remained there; receiving blows and dealing blows; showing a sturdy, positive force; mastering and employing all the methods of attack and defence allowed in such encounters, and supporting themselves with that fortitude and courage so necessary to the ring. Such combats are not to be commended, as they are usually mere tests of skill and endurance, entered into on the principles of the gambler, and they are introduced here for the sole purpose of showing the colored man as a positive force, yielding only to a superior degree of force of the same kind. The soldier stands for something far higher than the pugilist represents, although he has need of the same qualities of physical hardihood—contempt for suffering and coolness in the presence of danger, united with skill in the use of his weapons. The pugilist is his own general and never learns the high lessons of obedience; the soldier learns to subordinate himself to his commander, and to fight bravely and effectively under the direction of another.

The evolution of the Afro-American soldier was the work of a short period and suffered many interruptions. When the War of the Revolution broke out the colored man was a slave, knowing nothing of the spirit or the training of the soldier; before it closed several thousand colored men had entered the army and some had won distinction for gallantry. Less than forty years later, in the war of 1812, the black man again appeared to take his stand under the flag of independence. The War of Secession again witnessed the coming forth of the black soldier, this time in important numbers and performing heroic services on a grand scale, and under most discouraging circumstances, but with such success that he won a place in arms for all time. When the Civil War closed, the American black man had secured his standing as a soldier—the evolution was complete. Henceforth he was to be found an integral part of the Army of the United States.

The black man passed through the trying baptism of fire in the Sixties and came out of it a full-fledged soldier. His was worse than an impartial trial; it was a trial before a jury strongly biased against him; in the service of a government willing to allow him but half pay; and in the face of a foe denying him the rights belonging to civilized warfare. Yet against these odds, denied the dearest right of a soldier—the hope of promotion—scorned by his companions in arms, the Negro on more than two hundred and fifty battle-fields, demonstrated his courage and skill, and wrung from the American nation the right to bear arms. The barons were no more successful in their struggle with King John when they obtained Magna Charta than were the American Negroes with Prejudice, when they secured the national recognition of their right and fitness to hold a place in the Standing Army of the United States. The Afro-American soldier now takes his rank with America's best, and in appearance, skill, physique, manners, conduct and courage proves himself worthy of the position he holds. Combining in his person the harvested influences of three great continents, Europe, Africa and America, he stands up as the typical soldier of the Western World, the latest comer in the field of arms, but yielding his place in the line to none, and ever ready to defend his country and his flag against any and all foes.

The mission of this book is to make clear this evolution, giving the historical facts with as much detail as possible, and setting forth finally the portrait of this new soldier. That this is a prodigious task is too evident to need assertion—a task worthy the most lofty talents; and in essaying it I humbly confess to a sense of unfitness; yet the work lies before me and duty orders me to enter upon it. A Major General writes: "I wish you every success in producing a work important both historically and for the credit of a race far more deserving than the world has acknowledged." A Brigadier General who commanded a colored regiment in Cuba says to me most encouragingly: "You must allow me—for our intimate associations justify it—to write frankly. Your education, habits of thought, fairness of judgment and comprehension of the work you are to undertake, better fit you for writing such a history than any person within my acquaintance. Those noble men made the history at El Caney and San Juan; I believe you are the man to record it. May God help you to so set forth the deeds of that memorable first of July in front of Santiago that the world may see in its true light what those brave, intelligent colored men did."

Both these men fought through the Civil War and won distinction on fields of blood. To the devout prayer offered by one of them I heartily echo an Amen, and can only wish that in it all my friends might join, and that God would answer it in granting me power to do the work in such a way as to bring great good to the race and reflect some glory to Himself, in whose name the work is undertaken.

CHAPTER I.

SKETCH OF SOCIAL HISTORY.

The Importation of the Africans—Character of the Colored Population in 1860—Colored Population in British West Indian Possessions—Free Colored People of the South—Free Colored People of the North—Notes.

Professor DuBois, in his exhaustive work upon the "Suppression of the African Slave-Trade," has brought within comparatively narrow limits the great mass of facts bearing upon his subject, and in synopses and indices has presented all of the more important literature it has induced. In his Monograph, published as Volume II of the Harvard Historical Series, he has traced the rise of this nefarious traffic, especially with reference to the American colonies, exhibited the proportions to which it expanded, and the tenacity with which it held on to its purpose until it met its death in the fate of the ill-starred Southern Confederacy. Every step in his narrative is supported by references to unimpeachable authorities; and the scholarly Monograph bears high testimony to the author's earnest labor, painstaking research and unswerving fidelity. Should the present work stimulate inquiry beyond the scope herein set before the reader, he is most confidently referred to Professor Du Bois' book as containing a complete exposition of the development and overthrow of that awful crime.

It is from this work, however, that we shall obtain a nearer and clearer view of the African planted upon our shores. Negro slavery began at an early day in the North American Colonies; but up until the Revolution of 1688 the demand for slaves was mainly supplied from England, the slaves being white.* "It is probable," says Professor DuBois, "that about 25,000 slaves were brought to America each year between 1698 and 1707, and after 1713 it rose to perhaps 30,000 annually. "Before the Revolution the total exportation to America is variously estimated as between 40,000 and 100,000 each year." Something of the horrors of the "Middle Passage" may be shown by the records that out of 60,783 slaves shipped from Africa during the years 1680-88, 14,387, or nearly one-fourth of the entire number, perished at sea. In 1790 there were in the country nearly seven hundred thousand Africans, these having been introduced by installments from various heathen tribes. The importation of slaves continued with more or less success up until 1858, when the "Wanderer" landed her cargo of 500 in Georgia.

During the period from 1790 to the breaking out of the Civil War, shortly after the landing of the last cargo of slaves, the colored population, both slave and free, had arisen to about four million, and had undergone great modifications. The cargo of the "Wanderer" found themselves among strangers, even when trying to associate with those who in color and hair were like themselves. The slaves of 1860 differed greatly from the slaves of a hundred years earlier. They had lost the relics of that stern warlike spirit which prompted the Stono insurrection, the Denmark Vesey insurrection, and the Nat Turner insurrection, and had accepted their lot as slaves, hoping that through God, freedom would come to them some time in the happy future. Large numbers of them had become Christians through the teaching of godly white women, and at length through the evangelistic efforts of men and women of their

---

*Slave Trade—Carey.

own race. Independent religious organizations had been formed in the North, and large local churches with Negro pastors were in existence in the South when the "Wanderer" landed her cargo. There had been a steady increase in numbers, indicating that the physical well-being of the slave was not overlooked, and the slaves had greatly improved in character. Sales made in South Carolina between 1850 and 1860 show "boys," from 16 to 25 years of age, bringing from $900 to $1000; and "large sales" are reported showing an "average of $620 each," "Negro men bringing from $800 to $1000," and a "blacksmith" bringing $1425. The averages generally obtained were above $600. A sale of 109 Negroes in families is reported in the "Charleston Courier" in which the writer says: "Two or three families averaged from $1000 to $1100 for each individual." The same item states also that "C. G. Whitney sold two likely female house servants, one for $1000, the other for $1190." These cases are presented to illustrate the financial value of the American slave, and inferentially the progress he had made in acquiring the arts of modern civilization. Slaves had become blacksmiths, wheelwrights, carriage-makers, carpenters, bricklayers, tailors, bootmakers, founders and moulders, not to mention all the common labor performed by them. Slave women had become dressmakers, hairdressers, nurses and the best cooks to be found in the world. The slave-holders regarded themselves as the favored of mankind because of the competence and faithfulness of their slaves. The African spirit and character had disappeared, and in their place were coming into being the elements of a new character, existing in 1860 purely in a negative form. The slave had become an American. He was now a civilized slave, and had received his civilization from his masters. He had separated himself very far from his brother slave in St. Domingo. The Haytian Negro fought and won his freedom before he had been civilized in slavery, and hence has never passed over the same ground that his American fellow-servant has been compelled to traverse.

Beside the slaves in the South, there were also several thousand "free persons of color," as they were called, dwelling in such cities as Richmond, Va., Charleston, S. C., and New Orleans, La. Some of these had become quite wealthy and well-educated, forming a distinct class of the population. They were called creoles in Louisiana, and were accorded certain privileges, although laws were carefully enacted to keep alive the distinction between them and the whites. In Charleston the so-called colored people set themselves up as a class, prided themselves much upon their color and hair and in their sympathies joined almost wholly with the master class. Representatives of their class became slave-holders and were in full accord with the social policy of the country. Nevertheless their presence was an encouragement to the slave, and consequently was objected to by the slave-holder. The free colored man became more and more disliked in the South as the slave became more civilized. He was supposed by his example to contribute to the discontent of the slave, and laws were passed restricting his priveleges so as to induce him to leave. Between 1850 and 1860 this question reached a crisis and free colored people from the South were to be seen taking up their homes in the Northern States and in Canada. (Many of the people, especially from Charleston, carried with them all their belittling prejudices, and after years of sojourn under the sway of enlightened and liberal ideas, proved themselves still incapable of learning the new way or forgetting the old.)

There were, then, three very distinct classes of colored people in the country, to wit: The slave in the South, the free colored people of the South, and the free colored people of the North. These were also sub-divided into several smaller classes. Slaves were divided into field hands, house servants and city slaves. The free colored people of the South had their classes based usually on color; the free colored people of the North had their divisions caused by differences in religion, differences as to place of birth, and numerous family conceits. So that surveyed as a whole, it is extremely difficult to get anything like a complete social map of these four millions as they existed at the outbreak of the Civil War.

For a quarter of a century there had been a steady concentration of the slave population within the cotton and cane-growing region, the grain-growing States of Delaware, Maryland and Virginia having become to a considerable extent breeding farms. Particularly was this the case with the more intelligent and higher developed individual slaves who appeared near the border line. The master felt that such persons would soon make their escape by way of the "Underground Railroad" or otherwise, and hence in order to prevent a total loss, would follow the dictates of business prudence and sell his bright slave man to Georgia. The Maryland or Virginia slave who showed suspicious aspirations was usually checked by the threat, "I'll sell you to Georgia;" and if the threat did not produce the desired reformation it was not long before the ambitious slave found himself in the gang of that most despised and most despicable of all creatures, the Georgia slave-trader. Georgia and Canada were the two extremes of the slave's anticipation during the last decade of his experience. These stood as his earthly Heaven and Hell, the "Underground Railroad," with its agents, conducting to one, and the odious slave-trader, driving men, women and children, to the other. No Netherlander ever hated and feared the devil more thoroughly than did the slaves of the border States hate and fear these outrages on mankind, the kidnapping slave-traders of the cotton and cane regions. I say kidnapping, for I have myself seen persons in Georgia who had been kidnapped in Maryland. If the devil was ever incarnate, I think it safe to look for him among those who engaged in the slave-trade, whether in a foreign or domestic form.

Nothing is more striking in connection with the history of American Slavery than the conduct of Great Britain on the same subject. So inconsistent has this conduct been that it can be explained only by regarding England as a conglomerate of two elements nearly equal in strength, of directly opposite character, ruling alternately the affairs of the nation. As a slave-trader and slave-holder England was perhaps even worse than the United States. Under her rule the slave decreased in numbers, and remained a savage. In Jamaica, in St. Vincent, in British Guiana, in Barbadoes, in Trinidad and in Grenada, British slavery was far worse than American slavery. In these colonies "the slave was generally a barbarian, speaking an unknown tongue, and working with men like himself, in gangs with scarcely a chance for improvement." An economist says, had the slaves of the British colonics been as well fed, clothed, lodged, and otherwise cared for as were those of the United States, their number at emancipation would have reached from seventeen to twenty millions, whereas the actual number emancipated was only 660,000. Had the blacks of the United States experienced the same treatment as did those of the British colonies, 1860 would have found among us less than 150,000 colored persons. In the United States were found ten colored persons for every slave imported, while in the British colonies only one was found for every three imported. Hence the claim that the American Negro is a new race, built up on this soil, rests upon an ample supply of facts. The American slave was born in our civilization, fed upon good American food, housed and clothed on a civilized plan, taught the arts and language of civilization, acquired necessarily ideas of law and liberty, and by 1860 was well on the road toward fitness for freedom. No lessons therefore drawn from the emancipation of British slaves in the West Indies are of any direct value to us, inasmuch as British slavery was not like American slavery, the British freedman was in no sense the equal of the American freedman, and the circumstances surrounding the emancipation of the British slave had nothing of the inspiring and ennobling character with those connected with the breaking of the American Negro's chains. Yet, superior as the American Negro was as a slave, he was very far below the standard of American citizenship as subsequent events conclusively proved. The best form of slavery, even though it may lead toward fitness for freedom, can never be regarded as a fit school in which to graduate citizens of so magnificent an empire as the United States.

The slave of 1860 was perhaps, all things considered, the best slave the world had ever seen, if we except those who served the Hebrews under the Mosaic statutes. While there was no such thing among them as legal marriage or legitimate childhood, yet slave "families" were recognized even on the auction block, and after emancipation legal family life was erected generally upon relationships which had been formed in slavery. Bishop Gaines, himself born a slave of slave parents, says: "The Negro had no civil rights under the codes of the Southern States. It was often the case, it is true, that the marriage ceremony was performed, and thousands of couples regarded it, and observed it as of binding force, and were as true to each other as if they had been lawfully married." * * *

"The colored people generally," he says, "held their marriage (if such unauthorized union may be called marriage) sacred, even while they were slaves. Many instances will be recalled by the older people of the life-long fidelity which existed between the slave and his concubine" (Wife, T. G. S.) "...... the mother of his children. My own father and mother lived together over sixty years. I am the fourteenth child of that union, and I can truthfully affirm that no marriage, however made sacred by the sanction of law, was ever more congenial and beautiful. Thousands of like instances might be cited to the same effect. It will always be to the credit of the colored people that almost without exception, they adhered to their relations, illegal though they had been, and accepted gladly the new law which put the stamp of legitimacy upon their union and removed the brand of bastardy from the brows of their children."

Let us now sum up the qualifications that these people possessed in large degree, in order to determine their fitness for freedom, then so near at hand. They had acquired the English language, and the Christian religion, including the Christian idea of marriage, so entirely different in spirit and form from the African marriage. They had acquired the civilized methods of cooking their food, making and wearing clothes, sleeping in beds, and observing Sunday. They had acquired many of the useful arts and trades of civilization and had imbibed the tastes and feelings, to some extent, at least, of the country in which they lived. Becoming keen observers, shut out from books and newspapers, they listened attentively, learned more of law and politics than was generally supposed. They knew what the election of 1860 meant and were on tiptoe with expectation. Although the days of insurrection had passed and the slave of '59 was not ready to rise with the immortal John

Brown, he had not lost his desire for freedom. The steady march of escaping slaves guided by the North star, with the refrain:

> "I'm on my way to Canada,
> That cold but happy land;
> The dire effects of slavery
> I can no longer stand,"

proved that the desire to be free was becoming more extensive and absorbing as the slave advanced in intelligence.

It is necessary again to emphasize the fact that the American slaves were well formed and well developed physically, capable of enduring hard labor and of subsisting upon the plainest food. Their diet for years had been of the simplest sort, and they had been subjected to a system of regulations very much like those which are employed in the management of armies. They had an hour to go to bed and an hour to rise; left their homes only upon written "passes," and when abroad at night were often halted by the wandering patrol. "Run, nigger, run, the patrol get you," was a song of the slave children of South Carolina.

Strangers who saw for the first time these people as they came out of slavery in 1865 were usually impressed with their robust appearance, and a conference of ex-slaves, assembled soon after the war, introduced a resolution with the following declaration: "Whereas, Slavery has left us in possession of strong and healthy bodies." It is probable that at least a half-million of men of proper age could then have been found among the newly liberated capable of bearing arms. They were inured to the plain ration, to labor and fatigue, and to subordination, and had long been accustomed to working together under the immediate direction of foremen.

Two questions of importance naturally arose at this period: First, did the American slave understand the issue that had been before the country for more than a half-century and that was now dividing the nation in twain and marshalling for deadly strife these two opposing armies? Second, had he the courage necessary to take part in the struggle and help save the Union? It would be a strange thing to say, but nevertheless a thing entirely true, that many of the Negro slaves had a clearer perception of the real question at issue than did some of our most far-seeing statesmen, and a clearer vision of what would be the outcome of the war. While the great men of the North were striving to establish the doctrine that the coming war was merely to settle the question of Secession, the slave knew better. God had hid certain things from the wise and prudent and had revealed them unto babes. Lincoln, the wisest of all, was slow to see that the issue he himself had predicted was really at hand. As President, he declared for the preservation of the Union, with or without slavery, or even upon the terms which he had previously declared irreconcilable, "half slave and half free." The Negro slave saw in the outbreak of the war the death struggle of slavery. He knew that the real issue was slavery.

The masters were careful to keep from the knowledge of the slave the events as well as the causes of the war, but in spite of these efforts the slave's keen perception enabled him to read defeat in the dejected mien of his master, and victory in his exultation. To prevent the master's knowing what was going on in their thoughts, the slaves constructed curious codes among themselves. In one neighborhood freedom was always spoken of as "New Rice"; and many a poor slave woman sighed for the coming of New Rice in the hearing of those who imagined they knew the inmost thoughts of their bondwomen. Gleefully at times they would talk of the jollification they would make when the New Rice came. It was this clear vision, this strong hope, that sustained them during the trying days of the war and kept them back from insurrection. Bishop Gaines says: "Their prayers ascended for their deliverance, and their hearts yearned for the success of their friends. They fondly hoped for the hour of victory, when the night of slavery would end and the dawn of freedom appear. They often talked to each other of the progress of the war and conferred in secret as to what they might do to aid in the struggle. Worn out with long bondage, yearning for the boon of freedom, longing for the sun of liberty to rise, they kept their peace and left the result to God." Mr. Douglass, whom this same Bishop Gaines speaks of very inappropriately as a "half-breed," seemed able to grasp the feelings both of the slave and the freeman and said: "From the first, I for one, saw in this war the end of slavery, and truth requires me to say that my interest in the success of the North was largely due to this belief." Mr. Seward, the wise Secretary of State, had thought that the war would come and go without producing any change in the relation of master and slave; but the humble slave on the Georgia cotton plantation, or in the Carolina rice fields, knew that the booming of the guns of rebellion in Charleston was the opening note of the death knell of slavery. The slave undoubtedly understood the issue, and knew on which side liberty dwelt. Although thoroughly bred to slavery, and as contented and happy as he could be in his lot, he acted according to the injunction of the Apostle: "Art thou called being a servant, care not for it; but if thou mayest be made free, use it rather." The slaves tried to be contented, but they preferred freedom and knew which side to take when the time came for them to act.

Enough has been said to show that out of the African slave had been developed a thoroughly American slave, so well imbued with modern civilization and so well versed in American politics, as to be partially ready for citizenship. He had become law-abiding and order-loving, and possessed of an intelligent desire to be free. Whether he had within him the necessary moral elements to become a soldier the pages following will attempt to make known. He had the numbers, the physical strength and the intelligence. He could enter the strife with a sufficient comprehension of the issues involved to enable him to give to his own heart a reason for his action. Fitness for the soldier does not necessarily involve fitness for citizenship, but the actual discharge of the duties of the soldier in defence of the nation, entitles one to all common rights, to the nation's gratitude, and to the highest honors for which he is qualified.

In concluding this chapter I shall briefly return to the free colored people of the South that the reader may be able to properly estimate their importance as a separate element. Their influence upon the slave population was very slight, inasmuch as law and custom forbade the intercourse of these two classes.

According to the Census of 1860 there were in the slave-holding States altogether 261,918 free colored persons, 106,770 being mulattoes. In Charleston there were 887 free blacks and 2,554 mulattoes; in Mobile, 98 free blacks and 617 mulattoes; in New Orleans, 1,727 blacks and 7,357 mulattoes. As will be seen, nearly one-half of the entire number of free colored persons were mulattoes, while in the leading Southern cities seventy-five per cent. of the free colored people were put in this class. The percentage of mulatto slaves to the total slave population at that time was 10.41, and in the same cities which showed seventy-five per cent. of all the free colored persons mulattoes, the percentage of mulatto slaves was but 16.84. Mulatto in this classification includes all colored persons who are not put down as black.

In New Orleans the free mulattoes were generally French, having come into the Union with the Louisiana purchase, and among them were to be found wealthy slave-holders. They much resembled the class of mulattoes which obtained in St. Domingo at the beginning of the century, and had but little sympathy with the blacks, although they were the first to acquiesce in emancipation, some of them actually leading their own slaves into the army of liberation. It is possible, however, that they had not fully realized the trend of the war, inasmuch as New Orleans was excepted from the effects of the Proclamation. It is certain that the free colored people of that city made a tender of support to the Confederacy, although they were among the first to welcome the conquering "Yankees," and afterward fought with marked gallantry in the Union cause. The free mulattoes, or *browns,* as they called themselves, of Charleston, followed much the same course as their fellow classmen of New Orleans. Here, too, they had been exclusive and to some extent slave-holders, had tendered their services to the Confederacy, and had hastily come forward to welcome the conquerors. They were foremost among the colored people in wealth and intelligence, but their field of social operations had been so circumscribed that they had exerted but little influence in the work of Americanizing the slave. Separated from the slave by law and custom they did all in their power to separate themselves from him in thought and feeling. They drew the line against all blacks as mercilessly and senselessly as the most prejudiced of the whites and were duplicates of the whites placed on an intermediate plane. It was not unusual to find a Charleston brown filled with more prejudice toward the blacks than were the whites.

The colored people of the North in 1860 numbered 237,283,

*Census of 1860.

Pennsylvania having the largest number, 56,849; then came New York with 49,005; Ohio, 36,673; New Jersey, 25,318; Indiana, 11,428; Massachusetts, 9,602; Connecticut, 8,627; Illinois, 7,628; Michigan, 6,799; Rhode Island, 3,952; Maine, 1,327; Wisconsin, 1,171; Iowa, 1,069; Vermont, 709; Kansas, 625; New Hampshire, 494; Minnesota, 259; Oregon, 128.

Considerably more than one-half of this population was located within the States along the Atlantic Coast, viz.; Maine, New Hampshire, Massachusetts, Vermont, Connecticut, Rhode Island, New York, Pennsylvania and New Jersey. Here were to be found 154,883 free colored people. Pennsylvania, New York and New Jersey took the lead in this population, with Massachusetts and Connecticut coming next, while Maine, New Hampshire and Vermont had but few. The cities, Boston, New York and Philadelphia, were the largest cities of free colored people then in the North. In Boston there were 2,261; New York City, 12,574, while in Philadelphia there were 22,185

As early as 1787 the free colored people of Philadelphia, through two distinguished representatives, Absalom Jones and Richard Allen, "two men of the African race," as the chroniclers say, "saw the irreligious and uncivilized state" of the "people of their complexion," and finally concluded "that a society should be formed without regard to religious tenets, provided the persons lived an orderly and sober life," the purpose of the society being "to support one another in sickness and for the benefit of their widows and fatherless children." Accordingly a society was established, known as the Free African Society of Philadelphia, and on the 17th, 5th-mo., 1787, articles were published, including the following, which is inserted to show the breadth of the society's purpose:

"And we apprehend it to be necessary that the children of our deceased members be under the care of the Society, so far as to pay for the education of their children, if they cannot attend free school; also to put them out apprentices to suitable trades or places, if required."*

Shortly after this we read of "the African School for the free instruction of the black people," and in 1796, "The Evening Free School, held at the African Methodist Meeting House in Philadelphia" was reported as being "kept very orderly, the scholars behaving in a becoming manner, and their improvement beyond the teachers' expectations, their intellects appearing in every branch of learning to be equal to those of the fairest complexion." The name African, as the reader will notice, is used with reference to school, church, and individuals; although not to the complete exclusion of "colored people" and "people of color." These phrases seem to have been coined in the West Indies, and were there applied only to persons of mixed European and African descent. In the United States they never obtained such restricted use except in a very few localities. The practice of using African as a descriptive title of the free colored people of the North became very extensive and so continued up to the middle of the century. There were African societies, churches and schools in all the prominent centres of this population.

In 1843 one, Mr. P. Loveridge, Agent for Colored Schools of New York, wrote the editor of the African Methodist Magazine as follows:† "As to the name of your periodical, act as we did with the name of our schools—away with Africa. There are no Africans in your connection. Substitute colored for Afri-

*Outlines—Tanner.

†A. M. E. Magazine, 1843.

can and it will be, in my opinion, as it should be." The earnestness of the writer shows that the matter of parting with African was then a live question. The cool reply of the editor indicates how strong was the conservative element among the African people of '43. He says: "We are unable to see the reasonableness of the remarks. It is true we are not Africans, or natives born upon the soil of Africa, yet, as the descendants of that race, how can we better manifest that respect due to our fathers who begat us, than by the adoption of the term in our institutions, and inscribing it upon our public places of resort?" To this Mr. Loveridge rejoins in the following explanatory paragraph: "We who are engaged in the Public Schools in this city found upon examination of about 1500 children who attend our schools from year to year, not one African child among them. A suggestion was made that we petition the Public School Society to change the name African to Colored Schools. The gentlemen of that honorable body, perceiving our petition to be a logical one, acquiesced with us. Hence the adjective African (which does not apply to us) was blotted out and Colored substituted in its place. It is 'Public Schools for Colored Children.' We are Americans and expect American sympathies."

In 1816 the colored Methodists conceived the idea of organizing and evangelizing their race, and to this end a convention was called and assembled in Philadelphia of that year, composed of sixteen delegates, coming from Pennsylvania, Delaware, Maryland and New Jersey. The convention adopted a resolution that the people of Philadelphia, Baltimore and all other places who should unite with them, should become one body under the name and style of the African Methodist Episcopal Church. Similar action was taken by two other bodies of colored Methodists, one in New York, the other in Wilmington, Delaware, about the same time. The people were coming together and beginning to understand the value of organization. This was manifested in their religious, beneficial and educational associations that were springing up among them. In 1841 the African Methodist Magazine appeared, the first organ of religious communication and thought issued by the American colored people. It was published in Brooklyn, N. Y., Rev. George Hogarth being its editor.

There were papers published by the colored people prior to the appearance of the African Methodist Magazine, but these were individual enterprises. They were, however, indices of the thought of the race, and looking back upon them now, we may regard them as mile-stones set up along the line of march over which the people have come. New York, city and State, appears to have been the home of these early harbingers, and it was there that the earliest literary centre was established, corresponding to that centre of religious life and thought which had been earlier founded in Philadelphia. In 1827 the first newspaper published on this continent by colored men issued from its office in New York. It was called "Freedom's Journal," and had for its motto "Righteousness exalteth a nation." Its editors and proprietors were Messrs. Cornish & Russwurm. Its name was subsequently changed to the "Rights of All," Mr. Cornish probably retiring, and in 1830 it suspended, Mr. Russwurm going to Africa. Then followed "The Weekly Advocate," "The American," "The Colored American," "The Elevator," "The National Watchman," "The Clarion," "The Ram's Horn," "The North Star," "Frederick Douglass' Paper," and finally that crowning literary work of the race, "The Anglo-African."

"The Anglo-African" appeared in 1859, under the management of the strongest and most brilliant purely literary families the American Negro up to that time had produced. It was edited and published by Thomas Hamilton, and like all the important literary ventures of the race in those days, had its birth in New York. It came out in 1859 and continued through the war, and in 1865 went out of existence honorably, having its work well done. Its first volume, that of 1859, contains the ablest papers ever given to the public by the American Negro; and taken as a whole this volume is the proudest literary monument the race has as yet erected.

Reviewing the progress of the race in the North, we may say, the period of organized benevolence and united religious effort began before the close of the past century, Philadelphia being its place of origin; that the religious movement reached much broader and clearer standing about 1816, and in consequence there sprang up organizations comprehending the people of the whole country; that the religious movement advanced to a more intellectual stage when in 1841 the African Methodist Magazine appeared, since which time the organized religion of the American Negro has never been for any considerable time without its organs of communication. The journalistic period began in 1827, its centre being New York and the work of the journals almost wholly directed to two ends: the abolition of slavery, and the enfranchisement and political elevation of the free blacks. This work had reached its highest form in the Anglo-African, as that epoch of our national history came to its close in the slave-holders' war.

The titles of the newspapers indicate the opening and continuance of a period of anti-slavery agitation. Their columns were filled with arguments and appeals furnished by men who gave their whole souls to the work. It was a period of great mental activity on the part of the free colored people. They were discussing all probable methods of bettering their condition. It was the period that produced both writers and orators. In 1830 the first convention called by colored men to consider the general condition of the race and devise means to improve that condition, met in the city of Philadelphia. The history of this convention is so important that I append a full account of it as published in the Anglo-African nearly thirty years after the convention met. It was called through the efforts of Hezekiah Grice, of Baltimore, who afterwards emigrated to Hayti, and for many years followed there the occupation of carver and gilder and finally became Director of Public Works of the city of Port-au-Prince. While visiting that city years ago, I met a descendant of Mr. Grice, a lady of great personal beauty, charming manners, accomplished in the French language, but incapable of conversing at all in English.

The conventions, begun in 1830, continued to be held annually for a brief period, and then dropped into occasional and special gatherings. They did much good in the way of giving prominence to the colored orators and in stemming the tide of hostile sentiment by appealing to the country at large in language that reached many hearts.

The physical condition, so far as the health and strength of the free colored people were concerned, was good. Their mean age was the greatest of any element of our population, and their increase was about normal, or 1.50 per cent. annually. In the twenty years from 1840 to 1860 it had kept up this rate with hardly the slightest variation, while the increase of the free colored people of the South during the same period had been 1 per cent. annually.* The increase of persons of

*It is to be noted that in Maryland and Virginia an important number

mixed blood in the North did not necessarily imply laxity of morals, as the census compilers always delighted to say, but could be easily accounted for by the marriages occurring between persons of this class. I have seen more than fifty persons, all of mixed blood, descend from one couple, and these with the persons joined to them by marriages as they have come to marriageable age, amounted to over seventy souls—all in about a half century. That the slaves had, despite their fearful death rate, the manumissions and the escapes, increased twice as fast as the free colored people of the North, three times as fast as the freee colored people of the South, and faster than the white people with all the immigration of that period, can be accounted for only by the enormous birth rate of that people consequent upon their sad condition. Their increase was abnormal, and when properly viewed, proves too much.

There is no way of determining the general wealth of the colored people of the North at the period we are describing; but some light may be thrown upon their material condition from the consideration that they were supporting a few publications and building and supporting churches, and were holders of considerable real estate. In New York city, the thirteen thousand colored people paid taxes on nearly a million and a half in real estate, and had over a quarter million of dollars in the savings banks. It is probable that the twenty-five thousand in Philadelphia owned more in proportion than their brethren in New York, for they were then well represented in business in that city. There were the Fortens, Bowers, Casseys, Gordons, and later Stephen Smith, William Whipper and Videl, all of whom were men of wealth and business. There were nineteen churches owned and supported by colored

of white serving women married Negro slave men in the early days of these colonies.

people of Philadelphia, with a seating capacity of about 10,000 and valued at about $250,000.

*The schools set apart for colored children were very inferior and were often kept alive by great sacrifices on the part of the colored people themselves. Prior to the war and in many cases for some time afterward, the colored public schools were a disgrace to the country. A correspondent writing from Hollidaysburg, Pa., says, speaking of the school there: "The result of my inquiries here is that here, as in the majority of other places, the interest manifested for the colored man is more for political effect, and that those who prate the loudest about the moral elevation and political advancement of the colored man are the first to turn against him when he wants a friend." The correspondent then goes on to say that the school directors persist in employing teachers "totally incompetent." What the schools were in New York the report made by the New York Society for the promotion of Education among Colored Children to the Honorable Commissioners for examining into the condition of Common Schools in the City and County of New York, will show. Reverend Charles B. Ray, who was President of this Society, and Philip A. White, its Secretary, both continued to labor in the interest of education unto the close of their lives, Mr. White dying as a member of the School Board of the city of Brooklyn, and Mr. Ray bequeathing his library to Wilberforce University at his death.

In summing up the conditions which they have detailed in

*In 1835 there were six high schools, or schools for higher education, in the United States that admitted colored students on equal footing with others. These were: Oneida Institute, New York; Mount Pleasant, Amherst, Mass.; Canaan, N. H.; Western Reserve, Ohio; Gettysburg, Pa.; and "one in the city of Philadelphia of which Miss Buffam" was "principal." There was also one manual labor school in Madison County, N. Y., capable of accommodating eighteen students. It was founded by Gerrit Smith.

their report they say: "From a comparison of the school houses occupied by the colored children with the splendid, almost palatial edifices, with manifold comforts, conveniences and elegancies which make up the school houses for white children in the city of New York, it is clearly evident that the colored children are painfully neglected and positively degraded. Pent up in filthy neighborhoods, in old dilapidated buildings, they are held down to low associations and gloomy surroundings. * * * The undersigned enter their solemn protest against this unjust treatment of colored children. They believe with the experience of Massachusetts, and especially the recent experience of Boston before them, there is no sound reason why colored children shall be excluded from any of the common schools supported by taxes levied alike on whites and blacks, and governed by officers elected by the vote of colored as well as white voters."

This petition and remonstrance had its effect, for mainly through its influence within two years very great improvements were made in the condition of the New York colored schools.

For the especial benefit of those who erroneously think that the purpose of giving industrial education is a new thing in our land, as well as for general historical purposes, I call attention to the establishment of the Institute for Colored Youth in Philadelphia in 1842. This Institute was founded by the Society of Friends, and was supported in its early days and presumably still "by bequests and donations made by members of that Society." The objects of the Institute as set forth by its founders, fifty-seven years ago, are: "The education and improvement of colored youth of both sexes, to qualify them to act as teachers and instructors to their own people, either in the various branches of school learning or the mechanic arts and agriculture." Two years later the African Methodists purchased one hundred and eighty acres of land in eastern Ohio and established what was called the Union Seminary, on the manual labor plan. It did not succeed, but it lingered along, keeping alive the idea, until it was eclipsed by Wilberforce University, into which it was finally merged.

The anti-slavery fight carried on in the North, into which the colored men entered and became powerful leaders, aroused the race to a deep study of the whole subject of liberty and brought them in sympathy with all people who had either gained or were struggling for their liberties, and prompted them to investigate all countries offering to them freedom. No country was so well studied by them as Hayti, and from 1824 to 1860 there had been considerable emigration thither. Liberia, Central and South America and Canada were all considered under the thought of emigration. Thousands went to Hayti and to Canada, but the bulk preferred to remain here. They liked America, and had become so thoroughly in love with the doctrines of the Republic, so imbued with the pride of the nation's history, so inspired with hope in the nation's future, that they resolved to live and die on her soil. When the troublous times of 1860 came and white men were fleeing to Canada, colored men remained at their posts. They were ready to stand by the old flag and to take up arms for the Union, trusting that before the close of the strife the flag might have to them a new meaning. An impassioned colored orator had said of the flag: "Its stars were for the white man, and its stripes for the Negro, and it was very appropriate that the stripes should be red." The free Negro of the North was prepared in 1861 to support Abraham Lincoln with 40,000 as good American-born champions for universal liberty as the country could present.

NOTES.

A.

THE FIRST COLORED CONVENTION.

On the fifteenth day of September, 1830, there was held at Bethel Church, in the city of Philadelphia, the first convention of the colored people of these United States. It was an event of historical importance; and, whether we regard the times or the men of whom this assemblage was composed, we find matter for interesting and profitable consideration.

Emancipation had just taken place in New York, and had just been arrested in Virginia by the Nat Turner rebellion and Walker's pamphlet. Secret sessions of the legislatures of the several Southern States had been held to deliberate upon the production of a colored man who had coolly recommended to his fellow blacks the only solution to the slave question, which, after twenty-five years of arduous labor of the most hopeful and noble-hearted of the abolitionists, seems the forlorn hope of freedom to-day—insurrection and bloodshed. Great Britain was in the midst of that bloodless revolution which, two years afterwards, culminated in the passage of the Reform Bill, and thus prepared the joyous and generous state of the British heart which dictated the West India Emancipation Act. France was rejoicing in the not bloodless *trois jours de Juliet*. Indeed, the whole world seemd stirred up with a universal excitement, which, when contrasted with the universal panics of 1837 and 1857, leads one to regard as more than a philosophical speculation the doctrine of those who hold the life of mankind from the creation as but one life, beating with one heart, animated with one soul, tending to one destiny, although made up of millions upon millions of molecular lives, gifted with their infinite variety of attractions and repulsions, which regulate or crystallize them into evanescent substructures or organizations, which we call nationalities and empires and peoples and tribes, whose minute actions and reactions on each other are the histories which absorb our attention, whilst the grand universal life moves on beyond our ken, or only guessed at, as the astronomers shadow out movements of our solar system around or towards some distant unknown centre of attraction.

If the times of 1830 were eventful, there were among our people, as well as among other peoples, men equal to the occasion. We had giants in those days! There were Bishop Allen, the founder of the great Bethel connection of Methodists, combining in his person the fiery zeal of St. Francis Xavier with the skill and power of organizing of a Richelieu; the meek but equally efficient Rush (who yet remains with us in fulfilment of the Scripture), the father of the Zion Methodists; Paul, whose splendid presence and stately eloquence in the pulpit, and whose grand baptisms in the waters of Boston harbor are a living tradition in all New England; the saintly and sainted Peter Williams, whose views of the best means of our elevation are in triumphant activity to-day; William Hamilton, the thinker and actor, whose sparse specimens of eloquence we will one day place in gilded frames as rare and beautiful specimens of Etruscan art—William Hamilton, who, four years afterwards, during the New York riots, when met in the street, loaded down with iron missiles, and asked where he was going, replied, "To die on my threshold"; Watkins, of Baltimore; Frederick Hinton, with his polished eloquence; James Forten, the merchant prince; William Whipper, just essaying his youthful powers; Lewis Woodson and John Peck, of Pittsburg; Austin Steward, then of Rochester; Samuel E. Cornish, who had the distinguished honor of reasoning Gerrit Smith out of colonization, and of telling Henry Clay that he would never be president of anything higher than the American Colonization Society; Philip A. Bell, the born sabreur, who never feared the face of clay, and a hundred others, were the worthily leading spirits among the colored people.

And yet the idea of the first colored convention did not originate with any of these distinguished men; it came from a young man of Baltimore, then, and still, unknown to fame. Born in that city in 1801, he was in 1817 apprenticed to a man some two hundred miles off in the Southeast. Arriving at his field of labor, he worked hard nearly a week and received poor fare in return. One day, while at work near the house, the mistress came out and gave him a furious scolding, so furious, indeed, that her husband mildly interfered; she drove the latter away, and threatened to take the Baltimore out of the lad with cowhide, etc., etc. At this moment, to use his own expression, the lad became converted ,that is, he determined to be his own master as long as he lived. Early nightfall found him on his way to Baltimore which he reached after a severe journey which tested his energy and ingenuity to the utmost. At the age of twenty-three he was engaged in the summer time in supplying Baltimore with ice from his cart, and in winter in cutting up pork for Ellicotts' establishment. He must have been strong and swift with knife and cleaver, for in one day he cut up and dressed some four hundred and fifteen porkers.

In 1824 our young friend fell in with Benjamin Lundy, and in 1828-9, with William Lloyd Garrison, editors and publishers of the "Genius of Universal Emancipation," a radical anti-slavery paper, whose boldness would put the "National Era" to shame, printed and published in the slave State of Maryland. In 1829-30 the colored people of the free States were much excited on the subject of emigration; there had been an emigration to Hayti, and also to Canada, and some had been driven to Liberia by the severe laws and brutal conduct of the fermenters of colonization in Virginia and Maryland. In some districts of these States the disguised whites would enter the houses of free colored men at night, and take them out and give them from thirty to fifty lashes, to get them to consent to go to Liberia.

It was in the spring of 1830 that the young man we have sketched, Hezekiah Grice, conceived the plan of calling together a meeting or convention of colored men, in some place north of the Potomac, for the purpose of comparing views and of adopting a harmonious movement either of emigration or of determination to remain in the United States; convinced of the hopelessness of contending against the oppressions in the United States, living in the very depth of that oppression and wrong, his own views looked to Canada; but he held them subject to the decision of the majority of the convention which might assemble.

On the 2d of April, 1830, he addressed a written circular to prominent colored men in the free States, requesting their opinions on the necessity and propriety of holding such convention, and stated that if the opinions of a sufficient number warranted it, he would give time and place at which duly elected delegates might assemble. Four months passed away, and his spirit almost died within him, for he had not received a line from any one in reply. When he visited Mr. Garrison in his office, and stated his project, Mr. Garrison took up a copy of Walker's Appeal, and said, although it might be right, yet it was too early to have published such a book.

On the 11th of August, however, he received a sudden and peremptory order from Bishop Allen to come instantly to Philadelphia, about the emigration matter. He went, and found a meeting assembled to consider the conflicting reports on Canada of Messrs. Lewis and Dutton; at a subsequent meeting, held the next night, and near the adjournment, the Bishop called Mr. Grice aside and gave to him to read a printed circular, issued from New York City, strongly approving of Mr. Grice's plan of a convention, and signed by Peter Williams, Peter Vogelsang and Thomas L. Jinnings. The Bishop added, "My dear child, we must take some action immediately, or else these New Yorkers will get ahead of us." The Bishop left the meeting to attend a lecture on chemistry by Dr. Wells, of Baltimore. Mr. Grice introduced the subject of the convention; and a committee consisting of Bishop Allen, Benjamin Pascal, Cyrus Black, James Cornish and Junius C. Morel, were appointed to lay the matter before the colored people of Philadelphia. This committee, led, doubtless, by Bishop Allen, at once issued a call for a convention of the colored men of the United States, to be held in the city of Philadelphia on the 15th of September, 1830.

Mr. Grice returned to Baltimore rejoicing at the success of his project; but, in the same boat which bore him down the Chesapeake, he was accosted by Mr. Zollickoffer, a member of the Society of Friends, a Philadelphian, and a warm and tried friend of the blacks. Mr. Zollickoffer used arguments, and even entreaties, to dissuade Mr. Grice from holding the convention, pointing out the dangers and difficulties of the same should it succeed, and the deep injury it would do the cause in case of failure. Of course, it was reason and entreaty thrown away.

On the fifteenth of September, Mr. Grice again landed in Philadelphia, and in the fulness of his expectation asked every colored man he met about the convention; no one knew anything about it; the first man did not know the meaning of the word, and another man said, "Who ever heard of colored people holding a convention—convention, indeed!" Finally, reaching the place of meeting, he found, in solemn conclave, the five gentlemen who had constituted themselves delegates: with a warm welcome from Bishop Allen, Mr. Grice, who came with credentials from the people of Baltimore, was admitted as delegate. A little while after, Dr. Burton, of Philadelphia, dropped in, and demanded by what right the six gentlemen held their seats as members of the convention. On a hint from Bishop Allen, Mr. Pascal moved that Dr. Burton be elected an honorary member of the convention, which softened the Doctor. In half an hour, five or six grave, stern-looking men, members of the Zion Methodist body in Philadelphia, entered, and demanded to know by what right the members present held their seats and undertook to represent the colored people. Another hint from the Bishop, and it was moved that these gentlemen be elected honorary members. But the gentlemen would submit to no such thing, and would accept nothing short of full membership, which was granted them.

Among the delegates were Abraham Shadd, of Delaware; J. W. C. Pennington, of Brooklyn; Austin Steward, of Rochester; Horace Easton, of Boston, and —— Adams, of Utica.

The main subject of discussion was emigration to Canada; Junius C. Morel, chairman of a committee on that subject, presented a report, on which there was a two days' discussion; the point discussed was that the report stated that "the lands in Canada were synonymous with those of the Northern States." The word synonymous was objected to, and the word similar proposed in its stead. Mr. Morel, with great vigor and ingenuity, defended the report, but was finally voted down, and the word similar adopted. The convention recommended emigration to Canada, passed strong resolutions against the American Colonization Society, and at its adjournment appointed the next annual convention of the people of color to be held in Philadelphia, on the first Monday in June, 1831.

At the present day, when colored conventions are almost as frequent as church meetings, it is difficult to estimate the bold and daring spirit which inaugurated the Colored Convention of 1830. It was the right move, originating in the right quarter and at the right time. Glorious old Maryland, or, as one speaking in the view that climate grows the men, would say,—Maryland-Virginia region,—which has produced Benjamin Banneker, Nat. Turner, Frederick Douglass, the parents of Ira Aldridge, Henry Highland Garnett and Sam. Ringold Ward.

also produced the founder of colored conventions, Hezekiah Grice! At that time, in the prime of his young manhood, he must have presented the front of one equal to any fortune, able to achieve any undertaking. Standing six feet high, well-proportioned, of a dark bronze complexion, broad brow, and that stamp of features out of which the Greek sculptor would have delighted to mould the face of Vulcan—he was, to the fullest extent, a working man of such sort and magnetism as would lead his fellows where he listed.

In looking to the important results that grew out of this convention, the independence of thought and self-assertion of the black man are the most remarkable. Then, the union of purpose and union of strength which grew out of the acquaintanceship and mutual pledges of colored men from different States. Then, the subsequent conventions, where the great men we have already named, and others, appeared and took part in the discussions with manifestations of zeal, talent and ability, which attracted Garrison, the Tappans, Jocelyn and others of that noble host, who, drawing no small portion of their inspiration from their black brethren in bonds, did manfully fight in the days of anti-slavery which tried men's souls, and when, to be an abolitionist, was, to a large extent, to be a martyr.

We cannot help adding the thought that had these conventions of the colored people of the United States continued their annual sittings from 1830 until the present time, the result would doubtless have been greater general progress among our people themselves, a more united front to meet past and coming exigencies, and a profounder hold upon the public attention, and a deeper respect on the part of our enemies than we now can boast of. Looking at public opinion as it is, the living law of the land, and yet a malleable, ductile entity, which can be moulded, or at least affected, by the thoughts of any masses vigorously expressed, we should have become a power on earth, of greater strength and influence than in our present scattered and dwindled state we dare even dream of. The very announcement, "Thirtieth Annual Convention of the Colored People of the United States," would bear a majestic front. Our great gathering at Rochester in 1853, commanded not only public attention, but respect and admiration. Should we have such a gathering even now, once a year, not encumbered with elaborate plans of action, with too many wheels within wheels, we can yet regain much of the ground lost. The partial gathering at Boston, the other day, has already assumed its place in the public mind, and won its way into the calculations of the politicians.

Our readers will doubtless be glad to learn the subsequent history of Mr. Grice. He did not attend the second convention, but in the interval between the second and third he formed, in the city of Baltimore, a "Legal Rights Association," for the purpose of ascertaining the legal status of the colored man in the United States. It was entirely composed of colored men, among whom were Mr. Watkins (the colored Baltimorean), Mr. Deaver, and others. Mr. Grice called on William Wirt, and asked him "what he charged for his opinion on a given subject." "Fifty dollars." "Then, sir, I will give you fifty dollars if you will give me your opinion on the legal condition of a free colored man in these United States."

Mr. Wirt required the questions to be written out in proper form before he could answer them. Mr. Grice employed Tyson, who drew up a series of questions, based upon the Constitution of the United States, and relating to the rights and citizenship of the free black. He carried the questions to Mr. Wirt, who, glancing over them, said, "Really, sir, my position as an officer under the government renders it a delicate matter for me to answer these questions as they should be answered, but I'll tell you what to do: they should be answered, and by the best legal talent in the land; do you go to Philadelphia, and present my name to Horace Binney, and he will give you an answer satisfactory to you, and which will command the greatest respect throughout the land." Mr. Grice went to Philadelphia, and presented the questions and request to Horace Binney. This gentleman pleaded age and poor eyesight, but told Mr. Grice that if he would call on John Sargent he would get answers of requisite character and weight. He called on John Sargent, who promptly agreed to answer the questions if Mr. Binney would allow his name to be associated as an authority in the replies. Mr. Binney again declined, and so the matter fell through. This is what Mr. Grice terms his "Dred Scott case"and so it was.

He attended the convention of 1832, but by some informality, or a want of credentials, was not permitted to sit as full member!—Saul ejected from among the prophets!—Yet he was heard on the subject of rights, and the doctrine of "our rights," as well as the first colored convention, are due to the same man.

In 1832, chagrined at the colored people of the United States, he migrated to Hayti, where, until 1843, he pursued the business of carver and gilder. In the latter year he was appointed Director of Public Works in Port-au-Prince, which office he held until two years ago. He is also engaged in, and has wide knowledge of machinery and engineering. Every two or three years he visits New York, and is welcomed to the arcana of such men as James J. Mapes, the Bensons, Dunhams, and at the various works where steam and iron obey human ingenuity in our city. He is at present in this city, lodging at the house of the widow of his old friend and coadjutor, Thomas L. Jinnings, 133 Reade street. We have availed ourselves of his presence among us to glean from him the statements which we have imperfectly put together in this article.

We cannot dismiss this subject without the remark, of peculiar pertinence at this moment, that it would have been better for our people had Mr. Grice never left these United States. The twenty-seven years he has passed in Hayti, although not without their mark on the fortunes of that island, are yet with out such mark as he would have made in the land and upon the institutions among which he was born. So early as his thirty-second year, before he had reached his intellectual prime, he had inaugurated two of the leading ideas on which our people have since acted, conventions to consider and alleviate their grievances, and the struggle for legal rights. If he did such things in early youth, what might he not have done with the full force and bent of his matured intellect? And where, in the wide world, in what region, or under what sun, could he so effectually have labored to elevate the black man as on this soil and under American institutions?

So profoundly are we opposed to the favorite doctrine of the Puritans and their co-workers, the colonizationists—Ubi Libertas, ibi Patria—that we could almost beseech Divine Providence to reverse some past events and to fling back into the heart of Virginia and Maryland their Sam Wards, Highland Garnets, J. W. Penningtons, Frederick Douglasses, and the twenty thousand who now shout hosannas in Canada—and we would soon see some stirring in the direction of Ubi Patria, ibi Libertas.—Anglo-African Magazine, October, 1859.

B.

COMMUNICATION FROM THE NEW YORK SOCIETY FOR THE PROMOTION OF EDUCATION AMONG COLORED CHILDREN.

To the Honorable the Commissioners for examining into the condition of Common Schools in the City and County of New York.

The following statement in relation to the colored schools in said city and county is respectfully presented by the New York Society for the Promotion of Education among Colored Children:

1. The number of colored children in the city and county of New York (estimated in 1855, from the census of 1850), between the ages of 4 and 17 years........................ 3,000

a. Average attendance of colored children at public schools in 1855 913
Average attendance of colored children in corporate schools supported by school funds (Colored Orphan Asylum) 240
——— 1,153

b. Proportion of average attendance in public schools of colored children to whole number of same is as 1 to 2.60.

2. The number of white children in the city of New York in 1855 (estimated as above), between the ages of 4 and 17 years 159,000

a. Average attendance of white children in public schools in 1855 43,858
Average attendance of white children in corporate schools supported by public funds 2,826
——— 46,684

b. Proportion of average attendance of white children in public schools to whole number of same is as 1 to 3.40.

3. From these facts it appears that colored children attend the public schools (and schools supported by public funds in the city of New York) in the proportion of 1 to 2.60, and that the white children attend similar schools in said city in the proportion of 1 to 3.40; that is to say, nearly 25 per cent. more of colored children than of white children attend the public schools.

and schools supported by public funds in the city of New York.

4. The number of colored children attending private schools in the city of New York, 125.

a. The number of white children attending private schools in 1850, census gave 10,560, which number has since been increased by the establishment of Catholic parochial schools, estimated in 1856, 17,560.

b. The proportion of colored children attending private schools to white children attending same, is as 1 to 140.

c. But the average attendance of colored children in all schools is about the same as that of the white in proportion, that is to say, as many colored children attend the public schools as do whites attend both public and private schools, in proportion to the whole number of each class of children.

Locality, capability, etc., of colored schools.

1. The Board of Education, since its organization, has expended in sites and buildings for white schools $1,600,000.

b. The Board of Education has expended for sites and buildings for colored schools (addition to building leased 19 Thomas), $1,000.

c. The two schoolhouses in possession of the Board now used for colored children were assigned to same by the Old Public School Society.

2. The proportion of colored children to white children attending public schools is as 1 to 40.

a. The sum expended on school buildings and sites of colored and white schools by the Board of Education is as 1 to 1,600.

3. a. Schoolhouse No. 1, for colored children, is an old building, erected in 1820 by the New York Manumission Society as a school for colored children, in Mulberry street, in a poor but decent locality. It has two departments, one male and one female; it consists of two stories only, and has two small recitation rooms on each floor, but as primary as well as grammar children attend each department, much difficulty and confusion arises from the want of class room for the respective studies. The building covers only part of the lot, and as it is the best attended and among the best taught of the colored schools, a new and ample school building, erected in this place, would prove a great attraction, and could be amply filled by children.

b. Schoolhouse No. 2, erected in Laurens street more than twenty years ago for colored children by the Public School Society, is in one of the lowest and filthiest neighborhoods, and hence, although it has competent teachers in the male and female departments, and a separate primary department, the attendance has always been slender, and will be until the school is removed to a neighborhood where children may be sent without danger to their morals.

c. School No. 3, for colored children, in Yorkville, is an old building, is well attended, and deserves, in connection with Schoolhouse No. 4, in Harlem, a new building midway between the present localities.

d. Schoolhouse No. 5, for colored children, is an old building, leased at No. 19 Thomas street, a most degraded neighborhood, full of filth and vice; yet the attendance on this school, and the excellence of its teachers, earn for it the need of a new site and new building.

e. Schoolhouse No. 6, for colored children, is in Broadway, near 37th street, in a dwelling house leased and fitted up for a school, in which there is always four feet of water in the cellar. The attendance good. Some of the school officers have repeatedly promised a new buliding.

f. Primary school for colored children, No. 1, is in the basement of a church on 15th street, near 7th avenue, in a good location, but premises too small for the attendance; no recitation rooms, and is perforce both primary and grammar school, to the injury of the progress of all.

g. Primary schools for colored children, No. 2 and 3, are in the rear of church, in 2d street, near 6th avenue; the rooms are dark and cheerless, and without the needful facilities of sufficient recitation rooms, etc.

From a comparison of the schoolhouses with the splendid, almost palatial edifices, with manifold comforts, conveniences and elegancies which make up the schoolhouses for white children in the city of New York, it is evident that the colored children are painfully neglected and positively degraded. Pent up in filthy neighborhoods, in old and dilapidated buildings, they are held down to low associations and gloomy surroundings.

Yet Mr. Superintendent Kiddle, at a general examination of colored schools held in July last (for silver medals awarded by the society now addressing your honorable body) declared the reading and spelling equal to that of any schools in the city.

The undersigned enter their solemn protest against this unjust treatment of colored children. They believe with the experience of Massachusetts, and especially the recent experience of Boston before them, there is no sound reason why colored children shall be excluded from any of the common schools supported by taxes levied alike on whites and blacks, and governed by officers elected by the vote of colored as well as white voters.

But if in the judgment of your honorable body common schools are not thus common to all, then we earnestly pray you to recommend to the Legislature such action as shall cause the Board of Education of this city to erect at least two well-appointed modern grammar schools for colored children on suitable sites, in respectable localities, so that the attendance of colored children may be increased and their minds be elevated in like manner as the happy experience of the honorable Board of Education has been in the matter of white children.

In addition to the excellent impulse to colored youth which these new grammar schools would give, they will have the additional argument of actual economy; the children will be taught with far less expense in two such schoolhouses than in the half dozen hovels into which they are now driven. It is a costly piece of injustice which educates the white scholar in a palace at $10 per year and the colored pupil in a hovel at $17 or $18 per annum.

Taxes, etc., of colored population of the city.

No proposition can be more reasonable than that they who pay taxes for schools and schoolhouses should be provided with schools and schoolhouses. The colored population of this city, in proportion to their numbers, pay their full share of the general and therefore of the school taxes. There are about nine thousand adults of both sexes; of these over three thousand are householders, rent-payers, and therefore tax-payers, in that sense of the word in which owners make tax-payers of their poor tenants. The colored laboring man, with an income of $200 a year, who pays $72 per year for a room and bedroom, is really in proportion to his means a larger tax-payer than the millionaire whose tax rate is thousands of dollars.

But directly, also, do the colored people pay taxes. From examinations carefully made, the undersigned affirm that there are in the city at least 1,000 colored persons who own and pay taxes on real estate.

Taxed real estate in the city of New York owned by colored persons	$1,400,000
Untaxed by colored persons (churches)	250,000
Personal estate	710,000
Money in savings banks	1,121,000
	$3,481,000

These figures indicate that in proportion to their numbers, the colored population of this city pay a fair share of the school taxes, and that they have been most unjustly dealt with. Their money has been used to purchase sites and erect and fit up schoolhouses for white children, whilst their own children are driven into miserable edifices in disgraceful localities. Surely, the white population of the city are too able, too generous, too just, any longer to suffer this miserable robbing of their colored fellow-citizens for the benefit of white children.

Praying that your honorable commission will take due notice of these facts, and recommend such remedy as shall seem to you best,

We have the honor to be, in behalf of the New York Society for the Promotion of Education among Colored Citizens.

Most respectfully yours,

CHARLES B. RAY, President.

PHILIP A WHITE, Secretary.

New York City, December 28, 1857.

CHAPTER II.

AMERICAN NEGRO AND THE MILITARY SPIRIT.

Early Literature of Negro Soldiers—Negro Soldiers in the War of the Revolution—The War of 1812—Negro Insurrections—Negro Troops in the Civil War—Notes.

"Do you think I'll make a soldier?" is the opening line of one of those delightful spirituals, originating among the slaves in the far South. I first heard it sung in the Saint James Methodist Church, corner of Spring and Coming Streets, Charleston, South Carolina, immediately after the close of the war. It was sung by a vast congregation to a gentle, swinging air, with nothing of the martial about it, and was accompanied by a swaying of the body to the time of the music. Occasionally there would be the "curtesys" peculiar to the South Carolina slave of the low country, which consists in a stooping of the body by bending the knees only, the head remaining erect, a movement which takes the place of the bow among equals. The older ladies, with heads adorned with the ever-present Madras kerchief, often tied in the most becoming and tasteful manner, and faces aglow with an enthusiasm that bespoke a life within sustained by visions of spiritual things, would often be seen to shake hands and add a word of greeting and hope which would impart a charm and meaning to the singing far above what the humble words of the song without these accessories could convey. As the rich chorus of matchless voices poured out in perfect time and tune, "Rise, shine, and give God the glory," the thoughts of earthly freedom, of freedom from sin, and finally of freedom from the toils, cares and sorrows of earth to be baptized into the joys of heaven, all seemed to blend into the many colored but harmonious strain. The singing of the simple hearted trustful, emancipated slave! Shall we ever hear the like again on earth? Alas, that the high hopes and glowing prophecies of that auspicious hour have been so deferred that the hearts of millions have been made sick!

Of the songs that came out of slavery with these long suffering people, Colonel Higginson, who perhaps got nearer to them in sentiment than any other literary man not really of them, says: "Almost all their songs were thoroughly religious in their tone, however quaint their expression, and were in a minor key both as to words and music. The attitude is always the same, and, as a commentary on the life of the race, is infinitely pathetic. Nothing but patience for this life—nothing but triumph in the next. Sometimes the present predominates, sometimes the future; but the combination is always implied."

I do not know when this "soldier" song had its birth, but it may have sprung out of the perplexity of the slave's mind as he contemplated the raging conflict and saw himself drawn nearer and nearer to the field of strife. Whether in this song the "present predominates," and the query, therefore, has a strong primary reference to carnal weapons and to garments dyed in blood; whether the singer invites an opinion as to his fitness to engage in the war for Freedom—it may not be possible to determine. The "year of Jubilee," coming in the same song in connection with the purpose for which the singer is to be made a soldier, gives clearer illustration of that combination of the present and future which Mr. Higginson says was always present in the spirituals of that period, if it shows no more. When it is remembered that at that time Charleston was literally trodden under foot by black soldiers in bright uniforms, whose coming seemed to the colored people of that city like a dream too good to be true, it is not hard to believe that this song had much of the present in it, and owed its birth to the circumstances of war.

Singularly enough the song makes the Negro ask the exact question which had been asked about him from the earliest days of our history as a nation, a question which in some form confronts him still. The question, as the song has it, is not one of fact, but one of opinion. It is not: Will I make a soldier? but: Do you think I will make a soldier? It is one thing to "make a soldier," another thing to have men think so. The question of fact was settled a century ago; the question of opinion is still unsettled. The Negro soldier, hero of five hundred battlefields, with medals and honors resting upon his breast, with the endorsement of the highest military authority of the nation, with Port Hudson, El Caney and San Juan behind him, is still expected by too many to stand and await the verdict of thought, from persons who never did "think" he would make a soldier, and who never will think so. As well expect the excited animal of the ring to *think* in the presence of the red rag of the toreador as to expect *them* to think on the subject of the Negro soldier. They can curse, and rant, when they see the stalwart Negro in uniform, but it is too much to ask them to think. To them the Negro can be a fiend, a brute, but never a soldier.

To John G. Whittier and to William C. Nell are we indebted for the earliest recital of the heroic deeds of the colored American in the Wars of the Revolution and 1812. Whittier contributed an article on this subject to the "National Era" in 1847, and five or six years later Nell published his pamphlet on "Colored Patriots," a booklet recently reprinted by the African Methodist Episcopal Church. It is a useful contribution, showing as it does the rising and spreading abroad of that spirit which appreciates military effort and valor; and while recognizing the glory that came to American arms in the period described, honestly seeks to place some of that glory upon the deserving brow of a race then enslaved and despised. The book is unpretentious and aims to relate the facts in a straight-forward way, unaccompanied by any of the charms of tasteful presentation. Its author, however, is deserving our thanks, and the book marks an important stage in the development of the colored American. His mind was turning toward the creation of the soldier—the formation of armies.

There are other evidences that the mind of the colored man was at this time turning towards arms. In 1852 Doctor Pennington, one of the most learned colored men of his times, having received his Degree in Divinity from Heidelberg, delivered an address before a mass convention of colored citizens of Ohio, held in Cleveland, in which he spoke principally of the colored soldier. During the convention the "Cleveland Light Artillery" fired a salute, and on the platform were seated several veteran colored men, some of them, particularly Mr. John Julius, of Pittsburg, Pa., taking part in the speech-making. Mr. Nell says: "Within recent period several companies of colored men in New York city have enrolled themselves a la militaire," and quotes from the New York Tribune of August, 1852, as follows:

"COLORED SOLDIERS.—Among the many parades within a few days we noticed yesterday a soldierly-looking company of colored men, on their way homeward from a target or parade drill. They looked like men, handled their arms like men, and should occasion demand, we presume they would fight like men."

In Boston, New Haven, New Bedford and other places efforts were made during the decade from 1850 to 1860 to manifest this rising military spirit by appropriate organization, but the efforts were not always successful. In some cases the prejudices of the whites put every possible obstacle in the way of the colored young men who attempted to array themselves as soldiers.

The martial spirit is not foreign to the Negro character, as has been abundantly proved in both ancient and modern times. Williams, in his admirable history of the Negro as well as in his "Negro Troops in the Rebellion," has shown at considerable length that the Negro has been a soldier from earliest times, serving in large numbers in the Egyptian army long before the beginning of the Christian era. We know that without any great modification in character, runaway slaves developed excellent fighting qualities as Maroons, in Trinidad, British Guiana, St. Domingo and in Florida. But it was in Hayti that the unmixed Negro rose to the full dignity of a modern soldier, creating and leading armies, conducting and carrying on war, treating with enemies and receiving surrenders, complying fully with the rules of civilized warfare, and evolving finally a Toussaint, whose military genius his most bitter enemies were compelled to recognize—Toussaint, who to the high qualities of the soldier added also the higher qualities of statesmanship. With Napoleon, Cromwell and Washington, the three great commanders of modern times who have joined to high military talent eminent ability in the art of civil government, we must also class Toussaint L'Ouverteur, the black soldier of the Antilles. Thiers, the prejudiced attorney of Napoleon, declares nevertheless that Toussaint possessed wonderful talent for government, and the fact ever remains that under his benign rule all classes were pacified and

San Domingo was made to blossom as the rose. In the armies of Menelek, in the armies of France, in the armies of England, as well as in the organization of the Zulu and Kaffir tribes the Negro has shown himself a soldier. If the Afro-American should fail in this particular it will not be because of any lack of the military element in the African side of his character, or for any lack of "remorseless military audacity" in the original Negro, as the historian, Williams, expresses it.

In our own Revolutionary War, the Negro, then but partially civilized, and classed with "vagabonds," held everywhere as a slave, and everywhere distrusted, against protest and enactment, made his way into the patriot army, fighting side by side with his white compatriots from Lexington to Yorktown. On the morning of April 19th, 1775, when the British re-enforcements were preparing to leave Boston for Lexington, a Negro soldier who had served in the French war, commanded a small body of West Cambridge "exempts" and captured Lord Percy's supply train with its military escort and the officer in command. As a rule the Negro soldiers were distributed among the regiments, thirty or forty to a regiment, and did not serve in separate organizations. Bishop J. P. Campbell, of the African Methodist Church, was accustomed to say "both of my grandfathers served in the Revolutionary War." In Varnum's Brigade, however, there was a Negro regiment and of it Scribner's history, 1897, says, speaking of the battle of Rhode Island: "None behaved better than Greene's colored regiment, which three times repulsed the furious charges of veteran Hessians." Williams says: "The black regiment was one of three that prevented the enemy from turning the flank of the American army. These black troops were doubtless regarded as the weak spot of the line, but they were not."

The colony of Massachusetts alone furnished 67,907 men for the Revolutionary War, while all the colonies together south of Pennsylvania furnished but 50,493, hence the sentiment prevailing in Massachusetts would naturally be very powerful in determining any question pertaining to the army. When the country sprang to arms in response to that shot fired at Lexington, the echoes of which, poetically speaking, were heard around the world, the free Negroes of every Northern colony rallied with their white neighbors. They were in the fight at Lexington and at Bunker Hill, but when Washington came to take command of the army he soon gave orders that no Negroes should be enlisted. He was sustained in this position by a council of war and by a committee of conference in which were representatives from Rhode Island, Connecticut and Massachusetts, and it was agreed that Negroes be rejected altogether. The American Negro's persistency in pressing himself where he is not *wanted* but where he is *eminently needed* began right there. Within six weeks so many colored men applied for enlistment, and those that had been put out of the army raised such a clamor that Washington changed his policy, and the Negro, who of all America's population contended for the privilege of shouldering a gun to fight for American liberty, was allowed a place in the Continental Army, the first national army organized on this soil, ante-dating the national flag. The Negro soldier helped to evolve the national standard and was in the ranks of the fighting men over whom it first unfolded its broad stripes and glittering stars.

---

*"To the Honorable General Court of the Massachusetts Bay:

"The subscribers beg leave to report to your Honorable House, which we do in justice to the character of so brave a man, that, under our own observation, we declare that a Negro man called Salem Poor, of Col. Frye's regiment, Capt. Ames' company, in the late battle at Charlestown, be-

It is in place here to mention a legion of free mulattoes and blacks from the Island of St. Domingo, a full account of whose services is appended to this section, who fought under D'Estaing with great distinction in the siege of Savannah, their bravery at that time saving the patriot army from annihilation.

When the Revolutionary War had closed the brave black soldier who had fought to give to the world a new flag whose every star should be a star of hope to the oppressed, and whose trinity of colors should symbolize Liberty, Equalty and Fraternity, found his race, and in some instances himself personally, encased in a cruel and stubborn slavery. For the soldier himself special provision had been made in both Northern and Southern colonies, but it was not always hearty or effective. In October, 1783, the Virginia Legislature passed an act for the relief of certain slaves who had served in the army

---

haved like an experienced officer, as well as an excellent soldier. We would only beg leave to say, in the person of this said Negro centres a brave and gallant soldier. The reward due to so great and distinguished a character we submit to the Congress.

"Cambridge, Dec. 5, 1775."

These black soldiers, fresh from heathen lands, not out of slavery, proved themselves as worthy as the best. In the battle of Bunker Hill, where all were brave, two Negro soldiers so distinguished themselves that their names have come down to us garlanded with the tributes of their contemporaries. Peter Salem, until then a slave, a private in Colonel Nixon's regiment of Continentals, without orders fired deliberately upon Major Pitcairn as he was leading the assault of the British to what appeared certain victory. Everet in speaking "of Prescott, Putnam and Warren, the chiefs of the day," mentions in immediate connection "the colored man, Salem, who is reported to have shot the gallant Pitcairn as he mounted the parapet." What Salem Poor did is not set forth, but the following is the wreath of praise that surrounds his name:

Jona. Brewer, Col.
Thomas Nixon, Lt.-Col.
Wm. Precott, Col.
Ephm. Corey, Lieut.
Joseph Baker, Lieut.
Joshua Row, Lieut.
Jonas Richardson, Capt.
Eliphalet Bodwell, Sgt.
Josiah Foster, Lieut.
Ebenr. Varnum, 2d Lieut.
Wm. Hudson Ballard, Capt.
William Smith, Capt.
John Morton, Sergt. (?)
Richard Welsh, Lieut.

whose "former owners were trying to force to return to a state of servitude, contrary to the principles of justice and their solemn promise." The act provided that each and every slave who had enlisted "by the appointment and direction of his owner" and had "been received as a substitute for any free person whose duty or lot it was to serve" and who had served faithfully during the term of such enlistment, unless lawfully discharged earlier, should be fully and completely emancipated and should be held and deemed free in as full and ample manner as if each and every one of them were specially named in the act. The act, though apparently so fair on its face, and interlarded as it is with patriotic and moral phrases, is nevertheless very narrow and technical, liberating only those who enlisted by the appointment and direction of their owners, and who were accepted as substitutes, and who came out of the army with good discharges. It is not hard to see that even under this act many an ex-soldier might end his days in slavery. The Negro had joined in the fight for freedom and when victory is won finds himself a slave. He was both a slave and a soldier, too often, during the war; and now at its close may be both a veteran and a slave.

The second war with Great Britain broke out with an incident in which the Negro in the navy was especially conspicuous. The Chesapeake, an American war vessel was hailed, fired upon and forced to strike her colors, by the British. She was then boarded and searched and four persons taken from her decks, claimed as deserters from the English navy. Three of these were Negroes and one white. The Negroes were finally dismissed with a reprimand and the white man hanged. Five years later hostilities began on land and no opposition was manifested toward the employment of Negro soldiers. Laws were passed, especially in New York, authorizing the formation of regiments of blacks with white officers. It is remarkable that although the successful insurrection of St. Domingo was so recent, and many refugees from that country at that time were in the United States, and our country had also but lately come into possession of a large French element by the Louisiana purchase, there was no fear of a servile insurrection in this country. The free colored men of New Orleans, under the proclamation of the narrow-minded Jackson, rallied to the defence of that city and bore themselves with commendable valor in that useless battle. The war closed, however, and the glory of the Negro soldier who fought in it soon expired in the dismal gloom of a race-slavery becoming daily more wide-spread and hopeless.

John Brown's movement was military in character and contemplated the creation of an army of liberated slaves; but its early suppression prevented any display of Negro valor or genius. Its leader must ever receive the homage due those who are so moved by the woes of others as to overlook all considerations of policy and personal risk. As a plot for the destruction of life it fell far short of the Nat Turner insurrection which swept off fifty-seven persons within a few hours. In purpose the two episodes agree. They both aim at the liberation of the slave; both were led by fanatics, the reflex production of the cruelty of slavery, and both ended in the melancholy death of their heroic leaders. Turner's was the insurrection of the slave and was not free from the mad violence of revenge; Brown's was the insurrection of the friend of the slave, and was governed by the high and noble purpose of freedom. The insurrections of Denmark Vesey in South Carolina, in 1822, and of Nat Turner, in Virginia, in 1831, show conclusively that the Negro slave possessed the courage, the cunning, the secretiveness and the intelligence to fight for his freedom. These two attempts were sufficiently broad and intelligent, when taken into consideration with the enforced ignorance of the slave, to prove the Negro even in his forlorn condition capable of daring great things. Of the probable thousands who were engaged in the Denmark Vesey insurrection, only fifteen were convicted, and these died heroically without revealing anything connected with the plot. Forty-three years later I met the son of Denmark Vesey, who rejoiced in the efforts of his noble father, and regarded his death on the gallows as a holy sacrifice to the cause of freedom. Turner describes his fight as follows: "The white men, eighteen in number, approached us to about one hundred yards, when one of them fired, and I discovered about half of them retreating. I then ordered my men to fire and rush on them. The few remaining stood their ground until we approached within fifty yards, when they fired and retreated. We pursued and overtook some of them whom we thought we left dead. After pursuing them about two hundred yards, and rising a little hill, I discovered they were met by another party, and had halted and were reloading their guns. Thinking that those who retreated first and the party who fired on us at fifty or sixty yards distant had all only fallen back to meet others with ammunition, as I saw them reloading their guns, and more coming up than I saw at first, and several of my bravest men being wounded, the others became panic struck and scattered over the field. The white men pursued and fired on us several times. Hark had his horse shot under him, and I caught another for him that was running by me; five or six of my men were wounded, but none left on the field. Finding myself defeated here, I instantly determined to go through a private way and cross the Nottoway River at Cypress Bridge, three miles below Jerusalem, and attack that place in the rear, as I expected they would look for me on the other road, and I had a great desire to get there to procure arms and ammunition. After going a short distance in this private way, accompanied by about twenty men, I overtook two or three who told me the others were dispersed in every direction. After trying in vain to collect a sufficient force to proceed to Jerusalem, I determined to return, as I was sure they would make back to their old neighborhood, where they would rejoin me, make new recruits, and come down again. On my way back I called on Mrs. Thomas', Mrs. Spencer's and several other places. We stopped at Major Ridley's quarters for the night, and being joined by four of his men, with the recruits made since my defeat, we mustered now about forty strong.

After placing out sentinels, I lay down to sleep, but was quickly aroused by a great racket. Starting up I found some mounted and others in great confusion, one of the sentinels having given the alarm that we were about to be attacked. I ordered some to ride around and reconnoitre, and on their return the others being more alarmed, not knowing who they were, fled in different ways, so that I was reduced to about twenty again. With this I determined to attempt to recruit, and proceed on to rally in the neighborhood I had left."*

No one can read this account, which is thoroughly supported by contemporary testimony, without seeing in this poor misguided slave the elements of a vigorous captain. Failing in his efforts he made his escape and remained for two months in hiding in the vicinity of his pursuers. One concerned in his prosecution says: "It has been said that he was ignorant and cowardly and that his object was to murder and rob for the purpose of obtaining money to make his escape. It is notor-

*Confession of Nat Turner, Anglo-African Magazine, Vol. I, p. 338, 1859.

ious that he was never known to have a dollar in his life, to swear an oath, or drink a drop of spirits. As to his ignorance, he certainly never had the advantages of education, but he can read and write (it was taught him by his parents) and for natural intelligence and quickness of apprehension, is surpassed by few men I have ever seen. As to his being a coward, his reason as given for not resisting Mr. Phipps shows the decision of his character." *

The War of the Rebellion, now called the Civil War, effected the last and tremendous step in the transition of the American Negro from the position of a slave under the Republic to that of a soldier in its armies. Both under officers of his own race at Port Hudson and under white officers on a hundred battle-fields, the Negro in arms proved himself a worthy foeman against the bravest and sternest enemies that ever assailed our nation's flag, and a worthy comrade of the Union's best defenders. Thirty-six thousand eight hundred and forty-seven of them gave their lives in that awful conflict. The entire race on this continent and those of allied blood throughout the world are indebted to the soldier-historian, Honorable George W. Williams, for the eloquent story of their service in the Union Army, and for the presentation of the high testimonials to the valor and worthiness of the colored soldier as given by the highest military authority of the century. From Chapter XVI of his book, "Negro Troops in the Rebellion," the paragraphs appended at the close of this chapter are quoted.

---

†Ibid.

A.

HOW THE BLACK ST. DOMINGO LEGION SAVED THE PATRIOT ARMY IN THE SIEGE OF SAVANNAH, 1779.

The siege and attempted reduction of Savannah by the combined French and American forces is one of the events of our revolutionary war, upon which our historians care little to dwell. Because it reflects but little glory upon the American arms, and resulted so disastrously to the American cause, its important historic character and connections have been allowed to fade from general sight; and it stands in the ordinary school text-books, much as an affair of shame. The following, quoted from Barnes' History, is a fair sample of the way in which it is treated:

"French-American Attack on Savannah.—In September, D'Estaing joined Lincoln in besieging that city. After a severe bombardment, an unsuccessful assault was made, in which a thousand lives were lost. Count Pulaski was mortally wounded. The simple-hearted Sergeant Jasper died grasping the banner presented to his regiment at Fort Moultrie. D'Estaing refused to give further aid; thus again deserting the Americans when help was most needed."

From this brief sketch the reader is at liberty to infer that the attack was unwise if not fool-hardy; that the battle was unimportant; and that the conduct of Count D'Estaing immediately after the battle was unkind, if not unjust, to the Americans. While the paragraph does not pretend to tell the whole truth, what it does tell ought to be the truth; and this ought to be told in such a way as to give correct impressions. The attack upon Savannah was well-planned and thoroughly well considered; and it failed only because the works were so ably defended, chiefly by British regulars, under brave and skillful officers. In a remote way, which it is the purpose of this paper to trace, that sanguinary struggle had a wider bearing upon the progress of liberty in the Western World than any other one battle fought during the Revolution.

But first let us listen to the story of the battle itself. Colonel Campbell with a force of three thousand men, captured Savannah in December, 1778; and in the January following, General Prevost arrived, and by March had established a sort of civil government in Georgia, Savannah being the capital. In

April, the American general, Lincoln, feeble in more senses than one, perhaps, began a movement against Savannah by way of Augusta; but Prevost, aware of his purpose, crossed into South Carolina and attempted an attack upon Charleston. Finding the city too well defended, he contented himself with ravaging the plantations over a wide extent of adjacent country, and returned to Savannah laden with rich spoils, among which were included three thousand slaves, of whose labor he made good use later.

The patriots of the South now awaited in hope the coming of the French fleet; and on the first of September, Count D'Estaing appeared suddenly on the coast of Georgia with thirty-three sail, surprised and captured four British warships, and announced to the government of South Carolina his readiness to assist in the recapture of Savannah. He urged as a condition, however, that his ships should not be detained long off so dangerous a coast, as is was now the hurricane season, and there was neither harbor, road, nor offing for their protection.

By means of small vessels sent from Charleston he effected a landing in ten days, and four days thereafter, on the 16th, he summoned the garrison to surrender to the arms of France. Although this demand was made in the name of France for the plain reason that the American army was not yet upon the spot, the loyalists did not fail to make it a pretext for the accusation that the French were desirous of making conquests in the war on their own account. In the meantime Lincoln with the regular troops, was hurrying toward Savannah, and had issued orders for the militia to rendezvous at the same place; and the militia full of hope of a speedy, if not of a bloodless conquest, were entering upon this campaign with more than ordinary enthusiasm.

During the time that the fleet had been off the coast, and especially since the landing, the British had been very busy in putting the city in a high state of defence, and in making efforts to strengthen the garrison. Lieutenant-colonel Cruger, who had a small force at Sunbury, the last place in Georgia that had been captured by the British, and Lieutenant-colonel Maitland who was commanding a considerable force at Beaufort, were ordered to report in haste with their commands at Savannah. On the 16th, when the summons to surrender was received by Prevost, Maitland had not arrived, but was hourly expected. Prevost asked for a delay of twenty-four hours to consider the proposal, which delay was granted; and on that very evening, Maitland with his force arrived at Dawfuskie. Finding the river in the possession of the French, his course for a time seemed effectually cut off. By the merest chance he fell in with some Negro fishermen who informed him of a passage known as Wall's cut, through Scull's creek, navigable for small boats. A favoring tide and a dense fog enabled him to conduct his command unperceived by the French, through this route, and thus arrive in Savannah on the afternoon of the 17th, before the expiration of the twenty-four hours. General Prevost had gained his point; and now believing himself able to resist an assault, declined the summons to surrender. Two armed ships and four transports were sunk in the channel of the river below the city, and a boom in the same place laid entirely across the river; while several small boats were sunk above the town, thus rendering it impossible for the city to be approached by water.

On the day of the summons to surrender, although the works were otherwise well advanced, there were not ten cannon mounted in the lines of Savannah; but from that time until the day of assault, the men of the garrison, with the slaves they had captured, worked day and night to get the defences of the city in the highest state of excellence. Major Moncrief, chief of the engineers, is credited with placing in position more than eighty cannons in a short time after the call to surrender had been received.

The city itself at this time was but a mere village of frame buildings and unpaved streets. Viewed as facing its assailants, it was protected in its rear, or upon its north side, by the Savannah river; and on its west side by a thick swamp or morass, which communicated with the river above the city. The exposed sides were those of the east and south. These faced an open country which for several miles was entirely clear of woods. This exposed portion of the city was well protected by an unbroken line of defences extending from the river back to the swamp, the right and left extremes of the line consisting of strong redoubts, while the centre was made up of seamen's batteries in front, with impalements and traverses thrown up to protect the troops from the fire of the

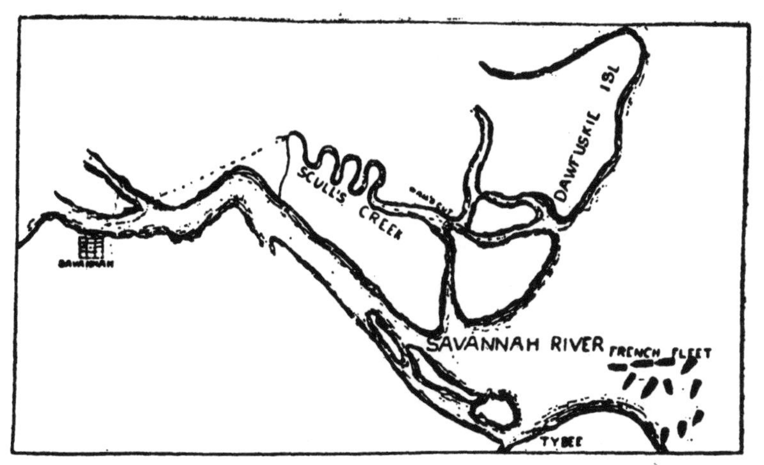

Savannah River.

besiegers. The whole extent of the works was faced with an ample abattis.

To be still more particular: there were three redoubts on the right of the line, and on the right of them quite near the swamp, was a sailor's battery of nine pounders, covered by a company of the British legion. The left redoubt of these three, was known as the Springhill redoubt; and proved to be the objective of the final assault. Between it and the centre, was another sailor's battery behind which were posted the grenadiers of the 60th regiment, with the marines which had been landed from the warships. On the left of the line near the river were two redoubts, strongly constructed, with a massy frame of green spongy wood, filled in with sand, and mounted with heavy cannon. The centre, or space between these groups of redoubts, was composed, as has been said, of lighter but nevertheless very effective works, and was strongly garrisoned.

Having thus scanned the works, let us now take a glance at the men who are to defend them. As all of the assaulting forces are not made up of Americans, so all of the defenders are not foreigners. The centre redoubt of the triplet on the right, was garrisoned by two companies of militia, with the North Carolina regiment to support them; Captains Roworth and Wylie, with the provincial corps of King's Rangers, were posted in the redoubt on the right; and Captain Tawse with his corps of provincial dragons, dismounted, in the left or Springhill redoubt, supported by the South Caroline regiment. The whole of this force on the right of the line, was under the command of the gallant Lieutenant-colonel Maitland; and it was this force that made the charge that barely failed of annihilating the American army. On the left of the line, the Georgia loyalists garrisoned one of those massy wooden sand-filled redoubts; while in the centre, cheek by jowl so to speak, with two battalions of the seventy-first regiment, and two regiments of Hessians, stood the New York Volunteers. All of these corps were ready to act as circumstances should require and to support any part of the line that might be attacked. The Negroes who worked on these defences were under the direction of Major Moncrief.

The French troops had landed below the city and were formed facing the British lines, with the river on their right.

On their left, later, assembled the American troops. The final dispositions were concluded by September 22nd, and were as follows: The American troops under Lincoln formed the left of the line, their left resting upon the swamp and the entire division facing the Springhill redoubt and her two sister defences; then came the division of M. de Noailles, composed of nine hundred men. D'Estaing's division of one thousand men beside the artillery, came next, and formed the centre of the French army. On D'Estaing's right was Count Dillon's division of nine hundred men; on the right of Dillon were the powder magazine, cattle depot, and a small field hospital; on the right of the depot and a little in advance, were Dejean's dragoons, numbering fifty men; upon the same alignment and to the right of the dragoons were Rouvrais' Volunteer Chasseurs, numbering seven hundred and fifty men; still further on to the right and two hundred yards in advance of Rouvrais, was Framais, comanding the Grenadier Volunteers, and two hundred men besides, his right resting upon the swampy wood that bordered the river, thus completely closing in the city on the land side. The frigate, La Truite, and two galleys, lay within cannon shot of the town, and with the aid of the armed store ship, La Bricole, and the frigate, La Chimere, effectually cut off all communication by water.

On the 23rd, both the French and the Americans opened their trenches; and on the 24th, a small detachment of the besieged made a sortie against the French. The attack was easily repulsed, but the French pursuing, approached so near the entrenchments of the enemy that they were fired upon and several were killed. On the night of the 27th another sortie was made which threw the besiegers into some confusion and caused the French and Americans to fire upon each other. Cannonading continued with but little result until October 8th.

The engineers were now of the opinion that a speedy reduction of the city could not be accomplished by regular approaches; and the naval officers were very anxious about the fleet, both because of the dangers to which it was exposed from the sea, and also because with so many men ashore it was in especial danger of being attacked and captured by British men-of-war. These representations agreeing altogether with D'Estaing's previously expressed wishes to leave the coast as soon as possible, induced that officer and General Lincoln to decide upon an attempt to storm the British works at once. It is quite probable that this had been the purpose as a last resort from the first. The preservation of the fleet was, however, the powerful factor in determining the time and character of the assault upon Savannah.

On the night of the eighth, Major L'Enfant, with a detachment attempted to set fire to the abattis in order to clear the way for the assault, but failed to through the dampness of the wood. The plan of the assault may be quite accurately obtained from the orders given to the American troops on the evening of the 8th by General Lincoln and from the inferences to be drawn from the events of the morning of the 9th as they are recorded in history. At least two of the historians who have left us accounts of the seige, Ramsey and McCall, were present at the time, and their accounts may be regarded as original authority. General Lincoln's orders were as follows:

"Evening Orders. By General Lincoln.
Watchword—Lewis.

"The soldiers will be immediately supplied with 40 rounds of cartridges, a spare flint, and have their arms in good order. The infantry destined for the attack of Savannah will be divided into two bodies; first composed of the light troops under the command of Colonel Laurens; the second, of the continental battalions and the first battalion of the Charleston militia, except the grenadiers, who are to join the light troops. The whole will parade at 1 o'clock, near the left of the line, and march by platoons. The guards of the camp will be formed of the invalids, and be charged to keep the fires as usual in camp.

"The cavalry under the command of Count Pulaski, will parade at the same time with the infantry and follow the left column of the French troops, precede the column of the American light troops; they will endeavor to penetrate the enemy's lines between the battery on the left of Springhill redoubt, and the next towards the river; having effected this, will pass to the left towards Yamacraw and secure such parties of the enemy as may be lodged in that quarter.

"The artillery will parade at the same time, follow the

French artillery, and remain with the corps de reserve until they receive further orders.

"The whole will be ready by the time appointed, with the utmost silence and punctuality; and be ready to march the instant Count Dillon and General Lincoln shall order.

"The light troops who are to follow the cavalry, will attempt to enter the redoubt on the left of the Springhill, by escalade if possible; if not by entrance into it, they are to be supported if necessary by the first South Carolina regiment; in the meantime the column will proceed with the lines to the left of the Springhill battery.

"The light troops having succeeded against the redoubt will proceed to the left and attempt the several works between that and the river.

"The column will move to the left of the French troops, taking care not to interfere with them.

"The light troops having carried the work towards the river will form on the left of the column.

"It is especially forbidden to fire a single gun before the redoubts are carried; or for any soldier to quit his rank to plunder without an order for that purpose; any who shall presume to transgress in either of these respects shall be reputed a disobeyer of military orders which is punishable with death.

"The militia of the first and second brigades, General Williamson's and the second battalion of the Charleston militia will parade immediately under the command of General Huger; after draughting five hundred of them the remainder of them will go into the trenches and put themselves under the commanding officer there; with the 500 he will march to the left of the enemy's line, remain as near them as he possibly can without being seen, until four o'clock in the morning, at which time the troops in the trenches will begin an attack upon the enemy; he will then advance and make his attack as near the river as possible; though this is only meant as a feint, yet should a favorable opportunity offer, he will improve it and push into the town.

"In case of a repulse after taking Springhill redoubt, the troops will retreat and rally in the rear of redoubt; if it cannot be effected that way, it must be attempted by the same route at which they entered.

"The second place of rallying (or the first if the redoubt should not be carried) will be at the Jews' burying-ground, where the reserve will be placed; if these two halts should not be effected, they will retire towards camp.

"The troops will carry in their hats a piece of white paper by which they will be distinguished."

General Huger with his five hundred militia, covered by the river swamp, crept quite close to the enemy's lines and delivered his attack as directed. Its purpose was to draw attention to that quarter and if possible cause a weakening of the strength in the left centre of the line. What its real effect was, there is now no means of knowing.

Count Dillon, who during the siege had been on D'Estaing's right, and who appears to have been second in command in the French army, in this assault was placed in command of a second attacking column. His purpose was to move to the right of General Huger, and keeping in the edge of the swamps along the river, steal past the enemy's batteries on the left, and attack him in the rear. Bancroft describes the results of his efforts as follows: "The column under Count Dillon, which was to have attacked the rear of the British lines, became entangled in a swamp of which it should only have skirted the edge was helplessly exposed to the British batteries and could not even be formed." Here were the two strong sand-filled redoubts, mounted with heavy cannon, and these may have been the batteries that stopped Dillon's column.

Count Pulaski with his two hundred brave cavalrymen, undertook his part in the deadly drama with ardor, and began that perilous ride which had for its object: "to penetrate the enemy's lines, between the battery on the left of the Springhill redoubt, and the next towards the river." Balch describes it as an attempt to "penetrate into the city by galloping between the redoubts." It was the anticipation of the Crimean "Charge of the Light Brigade;" only in this case, no one blundered; it was simply a desperate chance. Cannon were to the right, left, and front, and the heroic charge proved in vain; the noble Pole fell, banner* in hand, pierced with a

*The presentation of this banner by the Moravian Nuns of Bethlehem forms the text of the poem by Longfellow beginning—

When the dying flame of day
Through the chancel shot its ray,

mortal wound—another foreign martyr to our dearly bought freedom.

The cavalry dash having failed, that much of the general plan was blotted out. The feints may have been understood; it is said a sergeant of the Charleston Grenadiers deserted during the night of the 8th and gave the whole plan of the attack to General Prevost, so that he knew just where to strengthen his lines. The feints were effectually checked by the garrison on the left, twenty-eight of the Americans being killed: while Dillon's column was stopped by the batteries near the river. This state of affairs allowed the whole of Maitland's force to protect the Springhill redoubt and that part of the line which was most threatened. The Springhill redoubt, as has been stated, was occupied by the South Carolina regiment and a corps of dragoons. This circumstance may account for the fact, that while the three hundred and fifty Charleston militia occupied a most exposed position in the attacking column, only one man among them was killed and but six wounded. The battery on the left of this redoubt was garrisoned by grenadiers and marines.

The attacking column now advanced boldly, under the com-

Far the glimmering tapers shed
Faint light on the cowled head;
And the censer burning swung
Where, before the altar, hung
The crimson banner, that with prayer
Had been consecrated there,
And the nuns' sweet hymn was heard the while,
Sung low in the dim, mysterious aisle.
"Take thy banner! may it wave
Proudly o'er the good and brave;
When the battle's distant wail
Breaks the Sabbath of our vale,
When the cannon's music thrills
To the hearts of those lone hills,
When the spear in conflict shakes,
And the strong lance shivering breaks.

* * * * *

"Take thy banner! and if e'er
Thou should'st press the soldier's bier
And the muffled drum shall beat
To the tread of mournful feet,
Then the crimson flag shall be
Martial cloak and shroud for thee."
The warrior took that banner proud,
And it was his martial cloak and shroud.

mand of D'Estaing and Lincoln, the Americans consisting of six hundred continental troops and three hundred and fifty Charleston militia, being on the left, while the centre and right were made up of the French forces. They were met with so severe and steady a fire that the head of the column was soon thrown into confusion. They endured this fire for fifty-five minutes, returning it as best they could, although many of the men had no opportunity to fire at all. Two American standards and one French standard, were placed on the British works, but their bearers were instantly killed. It being found impossible to carry any part of the works, a general retreat was ordered. Of the six hundred continental troops, more than one-third had fallen, and about one-fifth of the French. The Charleston militia had not suffered, although they had bravely borne their part in the assault, and it had certainly been no fault of theirs if their brethren behind the embankments had not fired upon them. Count D'Estaing had received two wounds, one in the thigh, and being unable to move, was saved by the young naval lieutenant Truguet. Ramsey gives the losses of the battle as follows: French soldiers 760; officers 61; Americans 312; total 1133.

As the army began its retreat, Lieutenant-colonel Maitland with the grenadiers and marines, who were incorporated with the grenadiers, charged its rear with the purpose of accomplishing its annihilation. It was then that there occurred the most brilliant feat of the day, and one of the bravest ever performed by foreign troops in the American cause. In the army of D'Estaing was a legion of black and mulatto freedmen, known as Fontages Legion, commanded by Vicount de Fontages, a brave and experienced officer. The strength of this legion is given variously from six hundred to over eight hundred men. This legion met the fierce charge of Maitland and saved the retreating army.

In an official record prepared in Paris, now before me, are these words: "This legion saved the army at Savannah by bravely covering its retreat. Among the blacks who rendered signal services at that time were: Andre, Beauvais, Rigaud, Villatte, Beauregard, Lambert, who latterly became generals under the convention, including Henri Christophe, the future king of Haiti." This quotation is taken from a paper secured by the Honorable Richard Rush, our minister to

Paris in 1849, and is preserved in the Pennsylvania Historical Society. Henri Christophe received a dangerous gunshot wound in Savannah. Balch says in speaking of Fontages at Savannah: "He commanded there a legion of mulattoes, according to my manuscript, of more than eight hundred men, and saved the army after the useless assault on the fortifications, by bravely covering the retreat."

It was this legion that formed the connecting link between the siege of Savannah and the wide development of republican liberty on the Western continent, which followed early in the present century. In order to show this connection and the sequences, it will be necessary to sketch in brief the histor of this remarkable body of men, especially that of the prominent individuals who distinguished themselves at Savannah.

In 1779 the French colony of Saint Domingo was in a state of peace, the population then consisting of white slaveholders, mulatto and black freedmen (affranchis), and slaves. Count D'Estaing received orders to recruit men from Saint Domingo for the auxiliary army; and there being no question of color raised, received into the service a legion of colored freedmen. There had been for years a colored militia in Saint Domingo, and as early as 1716, the Marquis de Chateau Morant, then governor of the colony, made one Vincent the Captain-general of all the colored militia in the vicinity of the Cape. This Captain Vincent died in 1780 at the reputed age of 120 years. He was certainly of great age, for he had been in the siege of Carthegenia in 1697, was taken prisoner, afterwards liberated by exchange and presented to Louis XIV, and fought in the German war under Villars. Moreau de St. Mery, in his description of Vincent, incidentally mentions the Savannah expedition. He says: "I saw him (Vincent) the year preceding his death, recalling his ancient prowess to the men of color who were enrolling themselves for the expedition to Savannah; and showing in his descendants who were among the first to offer themselves, that he had transmitted his valor. Vincent, the good Captain Vincent, had a most pleasing countenance; and the contrast of his black skin with his white hair produced an effect that always commanded respect."

The Haytian historian, Enclus Robin, says when the call

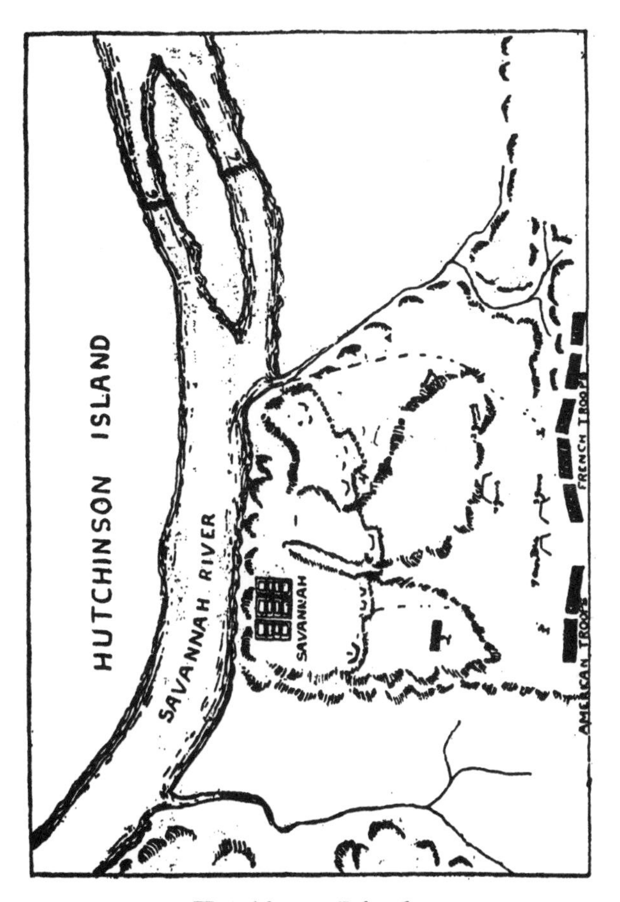

Hutchinson Island,

for volunteers reached Saint Domingo: "eight hundred young freedmen, blacks and mulattoes, offered themselves to take part in the expedition;" that they went and "fought valiantly; and returned to Saint Domingo covered with glory." Madiou, another Haytian historian of the highest respectability says: "A crowd of young men, black and colored, enlisted with the French troops and left for the continent They covered themselves with glory in the siege of Savannah, under the orders of Count D'Estaing."

What effect this experience had upon these volunteers may be inferred from their subsequent history. Robin says: "These men who contributed their mite toward American independence, had still their mothers and sisters in slavery; and they themselves were subject to humiliating discriminations. Should not France have expected from that very moment, that they would soon use in their own cause, those very arms which they had learned so well to use in the interests of others?" Madiou says: "On their return to Saint Domingo they demanded for their brothers the enjoyment of political rights." Beauvais went to Europe and served in the army of France; but returned to fight for liberty in Hayti, and was Captain-general in 1791; Rigaud, Lambert and Christophe wrote their names—not in the sand. These are the men who dared to stir Saint Domingo, under whose infleunce Hayti became the first country of the New World, after the United States, to throw off European rule. The connection between the siege of Savannah and the independence of Hayti is traced, both as to its spirit, and physically, through the black legion that on that occasion saved the American army. How this connection is traced to the republics of South America, I will allow a Haytian statesman and man of letters, honored both at home and abroad, to relate. I translate from a work published in Paris in 1885:

"The illustrious Bolivar, liberator and founder of five republics in South America, undertook in 1811 his great work of shaking off the yoke of Spain, and of securing the independence of those immense countries which swelled the pride of the catholic crown—but failed. Stripped of all resources he took flight and repaired to Jamaica, where he implored in vain of the governor of that island, the help of England. Almost in despair, and without means, he resolved to visit Hayti, and appeal to the generosity of the black Republic for the help necessary to again undertake that work of liberation which had gone to pieces in his hands. Never was there a more solemn hour for any man—and that man the representative of the destiny of South America! Could he hope for success? After the English, who had every interest in the destruction of Spanish colonial power, had treated him with so much indifference, could he hope that a new-born nation, weak, with microscopic territory, and still guarding anxiously its own ill-recognized independence, would risk itself in an enterprise hazardous as the one he represented? Full of doubt he came; but Petion gave him a most cordial welcome.

"Taking the precautions that a legitimate sentiment of prudence dictated at that delicate moment of our national existence, the government of Port-au-Prince put to the disposition of the hero of Boyaca and Carabobo, all the elements of which he had need—and Bolivar needed everything. Men, arms and money were generously given him. Petion did not wish to act openly for fear of compromising himself with the Spanish government; it was arranged that the men should embark secretly as volunteers; and that no mention of Hayti should ever be made in any official act of Venezuela."

Bolivar's first expedition with his Haytian volunteers was a failure; returning to the island he procured reinforcements and made a second descent which was brilliantly successful. Haytian arms, money and men turned Bolivar's disasters to victory; and the spirit of Western liberty marched on to the redemption of South America. The liberation of Mexico and all Central America, followed as a matter of course; and the ground was thus cleared for the practical application of that Continentalism enunciated in the Monroe doctrine.

The black men of the Antilles who fought in the siege of Savannah, enjoy unquestionably the proud historical distinction of being the physical conductors that bore away from our altars the sacred fire of liberty to rekindle it in their own land; and also of becoming the humble but important link that served to unite the Two Americas in the bond of enlightened independence.

T. G. STEWARD, U. S. A.

Note:—In the preparation of the above paper I have been greatly assisted by the Honorable L. J. Janvier, Charge d' affairs d' Haiti, in London; by Right Reverend James Theodore Holly, bishop of Hayti, and by Messrs. Charles and Frank Rudolph Steward of Harvard University. To all of these gentlemen my thanks are here expressed. T. G. S.

Paper read at the session of the Negro Academy, Washington, D. C., 1898.

B.

EXTRACTS FROM CHAPTER XVI "NEGRO TROOPS IN THE REBELLION"—WILLIAMS.

Adjutant-General Thomas in a letter to Senator Wilson, May 30, 1864, says: "Experience proves that they manage heavy guns very well. Their fighting qualities have also been fully tested a number of times, and I am yet to hear of the first case where they did not fully stand up to their work."

Major-General James G. Blunt writing of the battle of Honey Springs, Arkansas, said of Negro troops: "The Negroes (First Colored Regiment) were too much for the enemy, and let me here say that I never saw such fighting as was done by that Negro regiment. They fought like veterans, with a coolness and valor that is unsurpassed. They preserved their line perfect throughout the whole engagement, and although in the hottest of the fight, they never once faltered. Too much praise cannot be awarded them for their gallantry. The question that Negroes will fight is settled; besides, they make better soldiers in every respect than any troops I have ever had under my command."

General Thomas J. Morgan, speaking of the courage of Negro troops in the battle of Nashville, and its effect upon Major-General George H. Thomas, says: "Those who fell nearest the enemy's works were colored. General Thomas spoke very feelingly of the sight which met his eye as he rode over the field, and he confessed that the Negro had fully vindicated his bravery, and wiped from his mind the last vestige of prejudice and doubt."

CHAPTER III.

THE BLACK REGULARS OF THE ARMY OF INVASION IN THE SPANISH-AMERICAN WAR.

Organization of Negro Regiments in the Regular Army—First Move in the War—Chickamauga and Tampa—Note.

Altogether the colored soldiers in the Civil War took part and sustained casualties in two hundred and fifty-one different engagements and came out of the prolonged conflict with their character so well established that up to the present hour they have been able to hold an important place in the Regular Army of the United States. No regiment of colored troops in the service was more renowned at the close of the war or has secured a more advantageous position in the history of that period than the Fifty-fourth Massachusetts Regiment of Infantry. Recruited among the free colored people of the North, many of them coming from Ohio, it was remarkable for the intelligence and character of its men, and for the high purpose and noble bearing of its officers. Being granted but half the pay per month given to white soldiers, the regiment to a man, for eighteen months refused to receive one cent from the Government. This was a spectacle that the country could not longer stand. One thousand volunteers fighting the country's battles without any compensation rather than submit to a discrimination fatal to their manhood, aroused such a sentiment that Congress was compelled to put them on the pay-roll on equal footing with all other soldiers. By them the question of the black soldier's pay and rations was settled in the Army of the United States for all time. Every soldier, indeed every man in the army, except the chaplain, now draws the pay of his grade without regard to color, hair or race. By the time these lines reach the public eye it is to be hoped that even the chaplain will be lifted from his exceptional position and given the pay belonging to his rank as captain.

(February 2, 1901, the bill became a law giving chaplains the full pay of their grade.)

More than 185,000 blacks, all told, served in the army of the Union during the War of the Rebellion, and the losses from their ranks of men killed in battle were as heavy as from the white troops. Their bravery was everywhere recognized, and in the short time in which they were employed, several rose to commissions.

Perhaps the most notable act performed by a colored American during the war was the capture and delivery to the United States forces of the rebel steamer Planter, by Robert Smalls, of Charleston. Smalls was employed as pilot on the Planter, a rebel transport, and was entirely familiar with the harbors and inlets, of which there are many, on the South Atlantic coast. On May 13, 1862, the Planter came to her wharf in Charleston, and at night all the white officers went ashore, leaving a colored crew of eight men on board in charge of Smalls. Smalls hastily got his wife and three children on board, and at 2 o'clock on the morning of the 14th steamed out into the harbor, passing the Confederate forts by giving the proper signals, and when fairly out of reach, as daylight came, he ran up the Stars and Stripes and headed his course directly toward the Union fleet, into whose hands he soon surrendered himself and his ship. The act caused much favorable comment and Robert Smalls became quite a hero. His subsequent career has been in keeping with the high promise indicated by this bold dash for liberty, and his name has received additional lustre from gallant services performed in the war after, and in positions of distinguished honor and responsibility in civil life. The Planter, after being accepted by the United States, became a despatch boat, and Smalls demonstrating by skill and bravery his fitness for the position, was finally, as an act of imperative justice, made her commander.

With the close of the Revolutionary War the prejudice against a standing army was so great that the army was reduced to scarce six hundred men, and the Negro as a soldier dropped out of existence. When the War of 1812 closed sentiment with regard to the army had made but little advancement, and consequently no place in the service was left for Negro soldiers. In the navy the Negro still lingered, doing service in the lower grades, and keeping up the succession from the black heroes of '76 and 1812. When the War of the Rebellion closed the country had advanced so far as to see both the necessity of a standing army, and the fitness of the Negro to form a part of the army; and from this position it has never receded, and if the lessons of the Cuban campaign are rightly heeded, it is not likely to recede therefrom. The value of the Regular Army and of the Black Regular were both proven to an absolute demonstration in that thin line of blue that compelled the surrender of Santiago.

In July, 1866, Congress passed an act adding eight new regiments of infantry and four of calvary to the nineteen regiments of infantry and six of calvary of which those arms of the Regular Army were at that time composed, thus making the permanent establishment to consist of five regiments of artillery, twenty-seven of infantry, and ten of cavalry. Of the eight new infantry regiments to be formed, four were to be composed of colored men; and of the four proposed for the calvary arm, two were to be of colored men. The President was empowered by the act also to appoint a chaplain for each of the six regiments of colored troops. Under this law the Ninth and Tenth Cavalry Regiments were organized.

In 1869 the infantry suffered further reduction, and the four colored regiments organized under the law of 1866, numbered respectively the 38th, 39th, 40th and 41st, were consolidated into two regiments, and numbered the 24th and 25th—the 38th and 41st becoming the former, and the 39th and 40th the latter. Previous to this consolidation the numbers between the old 19th and the 38th, which was the lowest number borne by the new colored regiments, were filled in by dividing the old three batallion regiments in the service, and making of the second and third batallions of these regiments new regiments. The whole infantry arm, by the law of 1869, was compressed into twenty-five regiments, and in that condition the army remains to the present, to wit:* Ten regiments of cavalry, five of artillery and twenty-five of infantry.

The number of men in a company and the number of companies in a regiment have varied greatly within the past few months. Just previous to the breaking out of the war a regiment of infantry consisted of eight companies of about sixty men each, and two skeletonized companies and the band—the whole organization carrying about five hundred men; now a regiment of infantry consists of twelve companies of 106 men each and with the non-commissioned staff numbers twelve hundred and seventy-four men.

Since 1869, or for a period of thirty years, the colored American has been represented in the Regular Army by these four regiments and during this time these reigments have borne

*The army has been reorganized since. See Register.

more than their proportionate share in hard frontier service, including all sorts of Indian campaigning and much severe guard and fatigue duty. The men have conducted themselves so worthily as to receive from the highest military authority the credit of being among our best troops. General Miles and General Merritt,* with others who were active leaders in the Indian wars of the West, have been unstinting in their praise of the valor and skill of colored soldiers. They proved themselves not only good individual fighters, but in some instances non-commissioned officers exhibited marked coolness and ability in command.†

From 1869 to the beginning of the Hispano-American War there were in the Regular Army at some time, as commissioned officers, the following colored men, all from West Point, all serving with the cavalry, and none rising higher than first-lieutenant, viz: John H. Alexander, H. O. Flipper and Charles Young. H. O. Flipper was dismissed; Alexander died, and Young became major in the volunteer service, and was placed in command of the Ninth Battalion of Ohio Volunteers, discharging the duties of his position in such a manner as to command general satisfaction from his superior officers.**

These colored men while cadets at West Point endured hard-

*"My experience in this direction since the war is beyond that of any officer of my rank in the army. For ten years I had the honor of being lieutenant-colonel of the Ninth Cavalry, and during most of that service I commanded garrisons composed in part of the Ninth Cavalry and other organizations of cavalry and infantry. I have always found the colored race represented in the army obedient, intelligent and zealous in the discharge of duty, brave in battle, easily disciplined, and most efficient in the care of their horses, arms and equipments. The non-commissioned officers have habitually shown the qualities for control in their position which marked them as faithful and sensible in the discharge of their duties. I take pleasure in bearing witness as above in the interest of the race you represent." WESLEY MERRITT.

†See chapter on Colored Officers.

**Young is now captain in the Ninth Cavalry.—T. G. S.

ships disgraceful to their country, and when entering the army were not given that cordial welcome by their brother officers, becoming an "officer and gentleman," both to give and to receive. Of course there were some noble exceptions, and this class of officers seems to be steadily increasing, so that now it is no longer necessary, even on the ground of expediency, to strive to adhere to the rule of only white men for army officers. Of Alexander and Young it can be said they have acquitted themselves well, the former enjoying the confidence and esteem of his associates up to the time of his early death—an event which caused deep regret—and the latter so impressing the Governor of his State and the President as to secure for himself the responsible position which he, at the time of this writing, so worthily fills. Besides these line officers, five colored chaplains have been appointed, all of whom have served successfully, one, however, being dismissed by court-martial after many years of really meritorious service, an event to be regretted, but by no means without parallel.

Brief sketches of the history of these four colored regiments, as well as of the others, have been recently made by members of them and published in the Journal of the Military Service Institution and subsequently in a large and beautiful volume edited by Brigadier-General Theo. F. Rodenbough and Major William L. Haskin, published by the Institution and designated "The Army of the United States," a most valuable book of reference. From the sketches contained therein the following summary is given.

The Twenty-fourth Infantry was organized, as we have seen, from the 38th and 41st Regiments, these two regiments being at the time distributed in New Mexico, Louisiana and Texas, and the regiment remained in Texas from the time of its organization in 1869 until 1880. Its first Lieutenant-Colonel was

William R. Shafter. It was from this regiment and the Tenth Cavalry that the escort of Paymaster Wham was selected which made so brave a stand against a band of robbers that attacked the paymaster that several of them were given medals for distinguished gallantry, and others certificates of merit. The Twenty-fifth Infantry was organized in New Orleans out of the 39th, that was brought from North Carolina for that purpose, and the 40th, that was then in Louisiana. It was organized during the month of April, 1869, and early in 1870 moved to Texas, where it remained ten years. In 1880 it moved to the Department of Dakota and remained in the Northwest until it took the road for the Cuban war.

The Ninth Cavalry was organized in New Orleans during the winter of 1866-67. Its first Colonel was Edward Hatch and its first Lieutenant-Colonel Wesley Merritt. From 1867 to 1890 it was in almost constant Indian warfare, distinguishing itself by daring and hardihood. From 1890 to the opening of the Cuban war it remained in Utah and Nebraska, engaging in but one important campaign, that against hostile Sioux during the winter of 1890-91, in which, says the historian: "The regiment was the first in the field, in November, and the last to leave, late in the following March, after spending the winter, the latter part of which was terrible in its severity, under canvas."

The Tenth Calvary was organized under the same law as was the Ninth, and at the same time. Its place of rendezvous was Fort Leavenworth, Kansas, and its first Colonel, Benjamin H. Grierson. This regiment was the backbone of the Geronimo campaign force, and it finally succeeded in the capture of that wily warrior. The regiment remained in the Southwest until 1893, when it moved to Montana, and remained there until ordered to Chickamauga for the war.

These four regiments were finely officered, well drilled and well experienced in camp and field, particularly the cavalry regiments, and it was of them that General Merritt said: "I have always found them brave in battle." With such training and experience they were well fitted to take their place in that selected host of fighting men which afterwards became the Fifth Army Corps, placed under command of Major-General William R. Shafter, the first Lieutenant-Colonel of the Twenty-fourth Infantry.

When the news of the blowing up of our great battleship Maine, in the harbor of Havana, with the almost total loss of her crew, flashed over the country, carrying sadness to hundreds of homes, and arousing feelings of deepest indignation whether justly or unjustly, it was easy to predict that we should soon be involved in war with Spain. The Cuban question, already chronic, had by speeches of Senators Thurston and Proctor been brought to such a stage of aggravation that it needed only an incident to set the war element in motion. That incident was furnished by the destruction of the Maine. Thenceforth there was no power in the land sufficient to curb the rapidly swelling tide of popular hate, which manifested itself in the un-Christian but truly significant mottoes: "Remember the Maine," "Avenge the Maine," and "To hell with Spain." These were the outbreathings of popular fury, and they represented a spirit quite like that of the mob, which was not to be yielded to implicitly, but which could not be directly opposed.

The President did all in his power to stay this element of our population and to lead the country to a more befitting attitude. He and his advisers argued that Spain was to be resisted, and fought if necessary, not on account of the Maine, not in the spirit of revenge, but in the interest of humanity.

and upon principles sanctioned even by our holy religion. On behalf of the starving reconcentrados, and in aid of the noble Cuban patriot, we might justly arm and equip ourselves for the purpose of driving Spanish rule from the Western Hemisphere.

This view appealed to all lovers of freedom, to all true patriots, and to the Christian and philanthropist. It also afforded a superb opportunity for the old leaders in the South, who were not entirely relieved from the taint of secession, to come out and reconsecrate themselves to the country and her flag. Hence, Southern statesmen, who were utterly opposed to Negroes or colored men having any share in ruling at home, became very enthusiastic over the aspirations of the colored Cuban patriots and soldiers. The supporters, followers, and in a sense, devotees of Maceo and Gomez, were worthy of our aid. The same men, actuated by the same principles, in the Carolinas, in Louisiana or in Mississippi, would have been pronounced by the same authorities worthy of death.

The nation was, however, led into war simply to liberate Cuba from the iniquitous and cruel yoke of Spain, and to save thousands of impoverished Cubans from death by starvation. Great care was taken not to recognize the Cuban government in any form, and it seemed to be understood that we were to do the fighting both with our navy and our army, the Cubans being invited to co-operate with us, rather than that we should co-operate with them. We were to be the liberators and saviors of a people crushed to the very gates of death. Such was the platform upon which our nation stood before the world when the first orders went forth for the mobilization of its forces for war. It was a position worthy our history and character and gave to our national flag a prouder meaning than ever. Its character as the emblem of freedom shone out with awe-inspiring brilliancy amid the concourse of nations.

While there was such a clamor for war in the newspapers and in the public speeches of statesmen, both in and out of Congress, it is remarkable that the utmost serenity prevailed in the army. Officers and men were ready to fight if the stern necessity came, but they were not so eager for the death-game as were the numerous editors whose papers were getting out extras every half-hour. It was argued by the officers of rank that the Maine incident added nothing whatever to the Cuban question; that it did not involve the Spanish Government; that the whole subject might well be left to arbitration, and full respect should be given to Spain's disclaimer. It was also held that to rush into a war in order to prevent a few people from starving, might not relieve them, and at the same time would certainly cost the lives of many innocent men. Spain was revising her policy, and the benevolence of the United States would soon bring bread to the door of every needy Cuban. Such remarks and arguments as these were used by men who had fought through one war and were ready to fight through another if they must; but who were willing to go to any reasonable length to prevent it; and yet the men who used such arguments beforehand and manifested such a shrinking from carnage, are among those to whom the short Spanish War brought distinction and promotion. To their honor be it said that the war which gave them fresh laurels was in no sense brought about through their instigation.

As chaplain of the Twenty-fifth Infantry, stationed with the headquarters of the regiment at Fort Missoula, where we had been for ten years, the call for the war met me in the midst of my preparations for Easter service. One young man, then Private Thomas C. Butler, who was practicing a difficult solo for the occasion, before the year closed became a Second Lieutenant, having distinguished himself in battle; the janitor, who cared for my singing books, and who was my chief school teacher, Private French Payne, always polite and everywhere efficient, met his death from a Spanish bullet while on the reserve before bloody El Caney.

It was on a bright day during the latter part of March and near the close of the day as I was looking out of the front window of my quarters that I saw the trumpeter of the guard come out of the Adjutant's office with a dispatch in his hand and start on a brisk run toward the quarters of the Commanding Officer. I immediately divined what was in the wind, but kept quiet. In a few minutes "officers' call" was sounded, and all the officers of the post hastened to the administration building to learn the news.

When all were assembled the Commanding Officer desired to know of each company officer how much time he would need to have his company ready to move from the post to go to a permanet station elsewhere. and from all officers how much time they would require to have their families ready to quit the station. The answers generally were that all could be ready within a week. It was finally agreed, however, to ask for ten days.

Immediately the work of preparation began, although none knew where the regiment was to go. At this time the order, so far as it was understood at the garrison, was, that two companies were to go to Key West, Florida, and the other companies of the regiment to Dry Tortugas. One officer, Lieutenant V. A. Caldell, early saw through the haze and said: "It means that we will all eventually land in Cuba." While we were packing, rumors flew through the garrison, as indeed through the country, thick and fast, and our destination was changed three or four times a day. One hour we would be going to Key West, the next to St. Augustine, the next to Tortugas. In this confusion I asked an old frontier officer where he thought we would really go. Regarding himself as an indicator and always capable of seeing the amusing side of a subject, he replied: "I p'int toward Texas." Such was the state of uncertainty as to destination, and yet all the time the greatest activity prevailed in making ready for departure. Finally definite orders came that we were to store our furniture in the large gymnasium hall at the post and prepare to go in camp at Chickamauga Park, Georgia.

Our regiment was at the time stationed as follows: Headquarters, four companies and the band at Fort Missoula; two companies at Fort Harrison, near Helena, and two companies at Fort Assinniboine, all in Montana. The arrangements contemplated moving the regiment in two sections, one composed of the Missoula troops to go over the Northern Pacific Railroad, the other of the Fort Harrison and Fort Assinniboine troops to go over the Great Northern Railroad, all to arrive in St. Paul about the same time.

On the 10th of April, Easter Sunday, the battalion at Fort Missoula marched out of post quite early in the morning, and at Bitter Root Station took the cars for their long journey. Officers and men were all furnished sleeping accommodations on the train. Arriving in the city of Missoula, for the gratification of the citizens and perhaps to avoid strain on the bridge crossing the Missoula River, the men were disembarked from the train and marched through the principal streets to the depot, the citizens generally turning out to see them off. Many were the compliments paid officers and men by the good people of Missoula, none perhaps more pleasing than that furnished by a written testimonial to the regret experienced at the departure of the regiment, signed by all the ministers of the city.

As the Twenty-fifth was the first regiment to move in the

preparation for war, its progress from Montana to Chickamauga was a marked event, attracting the attention of both the daily and illustrated press. All along the route they were greeted with enthusiastic crowds, who fully believed the war with Spain had begun. In St. Paul, in Chicago, in Terre Haute, in Nashville, and in Chattanooga the crowds assembled to greet the black regulars who were first to bear forward the Starry Banner of Union and Freedom against a foreign foe. What could be more significant, or more fitting, than that these black soldiers, drilled up to the highest standard of modern warfare, cool, brave and confident, themselves a proof of American liberty, should be called first to the front in a war against oppression? Their martial tread and fearless bearing proclaimed what the better genius of our great government meant for all men dwelling beneath the protection of its honored flag.

As the Twenty-fifth Infantry was the first regiment to leave its station, so six companies of it were first to go into camp on the historic grounds of Chickamauga. Two companies were separated from the regiment at Chattanooga and forwarded to Key West where they took station under the command of Lieutenant-Colonel A. S. Daggett. The remaining six companies, under command of Colonel A. S. Burt, were conducted by General Boynton to a choice spot on the grounds, where they pitched camp, their tents being the first erected in that mobilization of troops which preceded the Cuban invasion, and theirs being really the first camp of the war.

Soon came the Ninth Cavalry, the Tenth Cavalry and the Twenty-fourth Infantry. While these were assembling there arrived on the ground also many white regiments, cavalry, artillery and infantry, and it was pleasing to see the fraternity that prevailed among black and white regulars. This was especially noticeable between the Twenty-fifth and Twelfth. In brigading the regiments no attention whatever was paid to the race or color of the men. The black infantry regiments were placed in two brigades, and the black cavalry likewise, and they can be followed through the fortunes of the war in the official records by their regimental numbers. During their stay in Chickamauga, and at Key West and Tampa, the Southern newspapers indulged in considerable malicious abuse of colored soldiers, and some people of this section made complaints of their conduct, but the previous good character of the regiments and the violent tone of the accusations, taken together with the well-known prejudices of the Southern people, prevented their complaints from having very great weight. The black soldiers held their place in the army chosen for the invasion of Cuba, and for that purpose were soon ordered to assemble in Tampa.

From the 10th of April, when the war movement began with the march of the Twenty-fifth Infantry out of its Montana stations, until June 14th, when the Army of Invasion cleared Tampa for Cuba—not quite two months—the whole energy of the War Department had been employed in preparing the army for the work before it. The beginning of the war is officially given as April 21st, from which time onward it was declared a state of war existed between Spain and the United States, but warlike movements on our side were begun fully ten days earlier, and begun with a grim definiteness that presaged much more than a practice march or spring manœuver.

After arriving at Chickamauga all heavy baggage was shipped away for storage, and all officers and men were required to reduce their field equipage to the minimum; the object being to have the least possible amount of luggage, in order that the greatest possible amount of fighting material might be carried. Even with all this preparation going on some officers were indulging the hope that the troops might remain in camps, perfecting themselves in drill, until September, or October, before they should be called upon to embark for Cuba. This, however, was not to be, and it is perhaps well that it was not, as the suffering and mortality in the home camps were almost equal to that endured by the troops in Cuba. The suffering at home, also, seemed more disheartening, because it appeared to be useless, and could not be charged to any important changes in conditions or climate. It was perhaps in the interest of humanity that this war, waged for humanity's sake, should have been pushed forward from its first step to its last, with the greatest possible dispatch, and that just enough men on our side were sent to the front, and no more. It is still a good saying that all is well that ends well.

The Chickamauga and Chattanooga National Military Park, the place where our troops assembled on their march to Cuba, beautiful by nature, especially in the full season of spring when the black soldiers arrived there, and adorned also by art, has, next to Gettysburg, the most prominent place among the historic battle-fields of the Civil War. As a park it was established by an act of Congress approved August 19, 1890, and contains seven thousand acres of rolling land, partly cleared and partly covered with oak and pine timber. Beautiful broad roads wind their way to all parts of the ground, along which are placed large tablets recording the events of those dreadful days in the autumn of 1863, when Americans faced Americans in bloody, determined strife. Monuments, judiciously placed, speak with a mute eloquence to the passer-by and tell of the valor displayed by some regiment or battery, or point to the spot where some lofty hero gave up his life. The whole park is a monument, however, and its definite purpose is to preserve and suitably mark "for historical and professional military study the fields of some of the most remarkable manœuvres and most brilliant fighting in the War of the Rebellion."

The battles commemorated by this great park are those of Chickamauga, fought on September 19-20, and the battles around Chattanooga, November 23-25, 1863. The battle of Chickamauga was fought by the Army of the Cumberland, commanded by Major-General W. S. Rosecrans, on the Union side, and the Army of Tennessee, commanded by General Braxton Bragg, on the side of the Confederates. The total effective strength of the Union forces in this battle was little less than 60,000 men, that of the Confederates about 70,000. The total Union loss was 16,179 men, a number about equal to the army led by Shafter against Santiago. Of the number reported as lost, 1,656 were killed, or as many as were lost in killed, wounded and missing in the Cuban campaign. The Confederate losses were 17,804, 2,389 being killed, making on both sides a total killed of 4,045, equivalent to the entire voting population of a city of over twenty thousand inhabitants. General Grant, who commanded the Union forces in the battles around Chattanooga, thus sums up the results: "In this battle the Union army numbered in round figures about 60,000 men; we lost 752 killed, 4,713 wounded and 350 captured or missing. The rebel loss was much greater in the aggregate, as we captured and sent North to be rationed there over 6,100 prisoners. Forty pieces of artillery, over seven thousand stand of small arms, many caissons, artillery wagons and baggage wagons fell into our hands. The probabilities are that our loss in killed was the heavier as we were the attacking party. The enemy reported his loss in killed at 361, but as he reported his missing at 4,146, while we held over 6,000 of them as prisoners, and there must have been hundreds, if not thousands, who deserted, but little reliance can be placed upon this report."

In the battle of Chickamauga, when "four-fifths of the Union Army had crumbled into wild confusion," and Rosecrans was intent only on saving the fragments, General Thomas, who had commanded the Federal left during the two days' conflict, and had borne the brunt of the fight, still held his position. To him General James A. Garfield reported. General Gordon Granger, without orders, brought up the reserves, and Thomas, replacing his lines, held the ground until nightfall, when he was joined by Sheridan. Bragg won and held the field, but Thomas effectually blocked his way to Chattanooga, securing to himself immediately the title of the "Rock of Chickamauga." His wonderful resolution stayed the tide of a victory dearly bought and actually won, and prevented the victors from grasping the object for which they had fought. In honor of this stubbborn valor, and in recognition of this high expression of American tenacity, the camp established in Chickamauga Park by the assembling army was called Camp George H. Thomas.

The stay of the colored regulars at Camp George H. Thomas was short, but it was long enough for certain newspapers of Chattanooga to give expression to their dislike to negro troops in general and to those in their proximity especially. The Washington Post, also, ever faithful to its unsavory trust, lent its influence to this work of defamation. The leading papers, however, both of Chattanooga and the South generally, spoke out in rather conciliatory and patronizing tones, and sought to restrain the people of their section from compromising their brilliant display of patriotism by contemptuous flings at the nation's true and tried soldiers.

The 24th Infantry and the 9th Cavalry soon left for Tampa,

Florida, whither they were followed by the 10th Cavalry and the 25th Infantry, thus bringing the entire colored element of the army together to prepare for embarkation. The work done at Tampa is thus described officially by Lieutenant-Colonel Daggett in general orders addressed to the 25th Infantry, which he at that time commanded. On August 11th, with headquarters near Santiago, after the great battles had been fought and won, he thus reviewed the work of the regiment: "Gathered from three different stations, many of you strangers to each other, you assembled as a regiment for the first time in more than twenty-eight years, on May 7, 1898, at Tampa, Florida. There you endeavored to solidify and prepare yourselves, as far as the oppressive weather would permit, for the work that appeared to be before you." What is here said of the 25th might have been said with equal propriety of all the regular troops assembled at Tampa.

In the meantime events were ripening with great rapidity. The historic "first gun" had been fired, and the United States made the first naval capture of the war on April 22, the coast trader Buena Ventura having surrendered to the American gunboat Nashville. On the same day the blockade of Cuban ports was declared and on the day following a call was issued for 125,000 volunteers. On May 20th the news that a Spanish fleet under command of Admiral Cervera had arrived at Santiago was officially confirmed, and a speedy movement to Cuba was determined upon.

Almost the entire Regular Army with several volunteer regiments were organized into an Army of Invasion and placed under the command of Major-General W. R. Shafter with orders to prepare immediately for embarkation, and on the 7th and 10th of June this army went on board the transports. For seven days the troops lay cooped up on the vessels awaiting orders to sail, a rumor having gained circulation that certain Spanish gunboats were hovering around in Cuban waters awaiting to swoop down upon the crowded transports. While the Army of Invasion was sweltering in the ships lying at anchor off Port Tampa, a small body of American marines made a landing at Guantanamo, and on June 12th fought the first battle between Americans and Spaniards on Cuban soil. In this first battle four Americans were killed. The next day, June 13th, General Shafter's army containing the four colored regiments, excepting those left behind to guard property, sailed for Cuba.*

The whole number of men and officers in the expedition, including those that came on transports from Mobile, amounted to about seventeen thousand men, loaded on twenty-seven transports. The colored regiments were assigned to brigades as follows: The Ninth Cavalry was joined with the Third and Sixth Cavalry and placed under command of Colonel Carrol; the Tenth Cavalry was joined with the Rough Riders and First Regular Cavalry and fell under the command of General Young; the Twenty-fourth Infantry was joined with the Ninth and Thirteenth Infantry and the brigade placed under command of Colonel Worth and assigned to the division commanded by General Kent, who, until his promotion as Brigadier-General of Volunteers, had been Colonel of the Twenty-fourth; the Twenty-fifth Infantry was joined with the First and Fourth Infantry and the brigade placed under command of Colonel Evans Miles, who had formerly been Major of the Twenty-fifth. All of the colored regiments were thus happily placed so that they should be in pleasant soldierly competition

*The colored regulars were embarked on the following named ships: The 9th Cavalry on the Miami, in company with the 6th Infantry; the 10th Cavalry on the Leona, in company with the 1st Cavalry; the 24th Infantry on the City of Washington, in company with one battalion of the 21st Infantry; the 25th infantry on board the Concho, in company with the 4th Infantry.

with the very best troops the country ever put in the field, and this arrangement at the start proves how strongly the black regular had entrenched himself in the confidence of our great commanders.

Thus sailed from Port Tampa the major part of our little army of trained and seasoned soldiers, representative of the skill and daring of the nation.* In physique, almost every man was an athlete, and while but few had seen actual war beyond an occasional skirmish with Indians, all excepting the few volunteers, had passed through a long process of training in the various details of marching, camping and fighting in their annual exercises in minor tactics. For the first time in history the nation is going abroad, by its army, to occupy the territory of a foreign foe, in a contest with a trans-Atlantic power. The unsuccessful invasions of Canada during the Revolutionary War and the War of 1812 can hardly be brought in comparison with this movement over sea. The departure of Decatur with his nine ships of war to the Barbary States had in view only the establishment of proper civil relations between those petty, half-civilized countries and the United States. The sailing of General Shafter's army was only one movement in a comprehensive war against the Kingdom of Spain. More than a month earlier Commodore Dewey, acting under orders, had destroyed a fleet of eleven war ships in the Philippines. The purpose of the war was to relieve the Cubans from an inhumane warfare with their mother country, and to restore to that unhappy island a stable government in harmony with the ideas of liberty and justice.

Up to the breaking out of the Spanish War the American policy with respect to Europe had been one of isolation. Some efforts had been made to consolidate the sentiment of the West-

*See Note, at the close of this chapter.

ern world, but it had never been successful. The fraternity of the American Republics and the attempted construction of a Pan-American policy had been thus far unfulfilled dreams. Canada was much nearer to the United States, geographically and socially, than even Mexico, although the latter is a republic. England, in Europe, was nearer than Brazil. The day came in 1898, when the United States could no longer remain in political seclusion nor bury herself in an impossible federation. Washington's advice against becoming involved in European affairs, as well as the direct corrollary of the Monroe Doctrine, were to be laid aside and the United States was to speak out to the world. The business of a European nation had become our business; in the face of all the world we resolved to invade her territory in the interest of humanity; to face about upon our own traditions and dare the opinions and arms of the trans-Atlantic world by openly launching upon the new policy of armed intervention in another's quarrel.

While the troops were mobilizing at Tampa preparatory to embarking for Cuba the question came up as to why there were no colored men in the artillery arm of the service, and the answer given by a Regular Army officer was, that the Negro had not brains enough for the management of heavy guns. It was a trifling assertion, of course, but at this period of the Negro's history it must not be allowed to pass unnoticed. We know that white men of all races and nationalities can serve big guns, and if the Negro cannot, it must be because of some marked difference between him and them. The officer said it was a difference in "brains," i. e., a mental difference. Just how the problem of aiming and firing a big gun differs from that of aiming and firing small arms is not so easily explained. In both, the questions of velocity, gravitation, wind and resistance are to be considered and these are largely settled by mechanism, the adjustment of which is readily learned; hence the assumption that a Negro cannot learn it is purely gratuitous. Several of the best rifle shots known on this continent are Negroes; and it was a Negro who summerized the whole philosophy of rifle shooting in the statement that it all consists in knowing *where* to aim, and *how* to pull—in knowing just what value to assign to gravitation, drift of the bullet and force of the wind, and then in being able to pull the trigger of the piece without disturbing the aim thus judiciously determined. This includes all there is in the final science and art of firing a rifle. If the Negro can thus master the revolver, the carbine and the rifle, why may he not master the field piece or siege gun?

But an ounce of fact in such things is worth more than many volumes of idle speculation, and it is remarkable that facts so recent, so numerous, and so near at hand, should escape the notice of those who question the Negro's ability to serve the artillery organizations. Negro artillery, both light and heavy, fought in fifteen battles in the Civil War with average effectiveness; and some of those who fought against them must either admit the value of the Negro artilleryman or acknowledge their own inefficiency. General Fitz-Hugh Lee failed to capture a Negro battery after making most vigorous attempts to that end. This attempt to raise a doubt as to the Negro's ability to serve in the artillery arm is akin to, and less excusable, than that other groundless assertion, that Negro officers cannot command troops, an assertion which in this country amounts to saying that the United States cannot command its army. Both of these assertions have been emphatically answered in fact, the former as shown above, and the latter as will be shown later in this volume. These assertions are only temporary covers, behind which discomfitted and retreating prejudice is able to make a brief stand, while the black hero of five hundred battle-fields, marches proudly by, disdaining to lower his gun to fire a shot on a foe so unworthy. When the Second Massachusetts Volunteers sent up their hearty cheers of welcome to the gallant old Twenty-fifth, as that solid column fresh from El Caney swung past its camp, I remarked to Sergeant Harris, of the Twenty-fifth: "Those men think you are soldiers." "They know we are soldiers," was his reply. When the people of this country, like the members of that Massachusetts regiment, come to know that its black men in uniform are soldiers, plain soldiers, with the same interests and feelings as other soldiers, of as much value to the government and entitled from it to the same attention and rewards, then a great step toward the solution of the prodigious problem now confronting us will have been taken.

Note.—"I had often heard that the physique of the men of our regular army was very remarkable, but the first time I saw any large body of them, which was at Tampa, they surpassed my highest expectations. It is not, however, to be wondered at that, for every recruit who is accepted, on the average thirty-four are rejected, and that, of course, the men who present themselves to the recruiting officer already represent a physical 'elite'; but it was very pleasant to see and be assured, as I was at Tampa, by the evidences of my own eyes and the tape measure, that there is not a guard regiment of either the Russian, German or English army, of whose remarkable physique we have heard so much, that can compare physically, not with the best of our men, but simply with the average of the men of our regular army."—Bonsal.

CHAPTER IV.

BRIEF SKETCH OF SPANISH HISTORY.

The following brief sketch of Spain, its era of greatness, the causes leading thereto, and the reasons for its rapid decline, will be of interest to the reader at this point in the narrative, as it will bring into view the other side of the impending conflict:

Spain, the first in rank among the second-rate powers of Europe, by reason of her possessions in the West Indies, especially Cuba, may be regarded as quite a near neighbor, and because of her connection with the discovery and settlement of the continent, as well as the commanding part she at one time played in the world's politics, her history cannot but awaken within the breasts of Americans a most lively interest.

As a geographical and political fact, Spain dates from the earliest times, and the Spanish people gather within themselves the blood and the traditions of the three great continents of the Old World—Europe, Asia and Africa—united to produce the mighty Spaniard of the 15th and 16th centuries. It would be an interesting subject for the anthropologist to trace the construction of that people who are so often spoken of as possessing the pure blood of Castile, and as the facts should be brought to view, another proud fiction would dissipate in thin air, as we should see the Spaniard arising to take his place among the most mixed of mankind.

The Spain that we are considering now is the Spain that gradually emerged from a chaos of conflicting elements into the unity of a Christian nation. The dismal war between creeds gave way to the greater conflict between religions, when Cross and Crescent contended for supremacy, and this too had passed. The four stalwart Christian provinces of Leon, Castile, Aragon and Navarre had become the four pillars of support to a national throne and Ferdinand and Isabella were reigning. Spain has now apparently passed the narrows and is crossing the bar with prow set toward the open sea. She ends her war with the Moors at the same time that England ends her wars of the Roses, and the battle of Bosworth's field may be classed with the capitulation of Granada. Both nations confront a future of about equal promise and may be rated as on equal footing, as this new era of the world opens to view.

What was this new era? Printing had been invented, commerce had arisen, gunpowder had come into use, the feudal system was passing, royal authority had become paramount, and Spain was giving to the world its first lessons in what was early stigmatized as the "knavish calling of diplomacy."

Now began the halcyon days of Spain, and what a breed of men she produced! Read the story of their conquests in Mexico and Peru, as told with so much skill and taste by our own Prescott; or read of the grandeur of her national character, and the wonderful valor of her troops, and the almost marvelous skill of her Alexander of Parma, and her Spinola, as described by our great Motley, and you will see something of the moral and national glory of that Spain which under Charles V and Philip II awed the world into respectful silence.

Who but men of iron, under a commander of steel, could have conducted to a successful issue the awful siege of Antwerp, and by a discipline more dreadful than death, kept for so many years, armed control of the country of the brave

Netherlanders? A Farnese was there, who could support and command an army, carry Philip and his puerile idosyncrasies upon his back and meet the fury of an outraged people who were fighting on their own soil for all that man holds dear. Never was wretched cause so ably led, never were such splendid talents so unworthily employed.

Alexander of Parma, Cortez, the Pizarros, were representatives of that form of human character that Spain especially developed. Skill and daring were brought out in dazzling splendor, and success followed their movements. Take a brief survey of the Empire under Charles V: Himself Emperor of Germany; his son married to the Queen of England; Turkey repulsed; France humbled, and all Europe practically within his grasp. And what was Spain outside of Europe? In America she possessed territory covering sixty degrees of latitude, owning Mexico, Central America, Venezuela, New Granada, Peru and Chili, with vast parts of North America, and the islands of Cuba, Jamaica and St. Domingo. In Africa and Asia she had large possessions—in a word, the energies of the world were at her feet. The silver and gold of America, the manufactures and commerce of the Netherlands, combined to make her the richest of nations.

The limits of the present purpose do not permit an exhaustive presentation of her material strength in detail, nor are the means at hand for making such an exhibit. We must be content with a general picture, quoted directly from Motley. He says:

"Look at the broad magnificent Spanish Peninsula, stretching across eight degrees of latitude and ten of longtitude, commanding the Atlantic and the Mediterranean, with a genial climate, warmed in winter by the vast furnace of Africa, and protected from the scorching heats of summer by shady mountain and forest, and temperate breezes from either ocean. A generous southern territory, flowing with oil and wine, and all the richest gifts of a bountiful nature—splendid cities—the new and daily expanding Madrid, rich in the trophies of the most artistic period of the modern world; Cadiz, as populous at that day as London, seated by the straits where the ancient and modern systems of traffic were blending like the mingling of the two oceans; Granada, the ancient, wealthy seat of the fallen Moors; Toledo, Valladolid, and Lisbon, chief city of the recently conquered kingdom of Portugal, counting with its suburbs a larger population than any city excepting Paris, in Europe, the mother of distant colonies, and the capital of the rapidly-developing traffic with both the Indies—these were some of the treasures of Spain herself. But she possessed Sicily also, the better portion of Italy, and important dependencies in Africa, while the famous maritime discoveries of the age had all enured to her aggrandizement. The world seemed suddenly to have expanded its wings from East to West, only to bear the fortunate Spanish Empire to the most dizzy heights of wealth and power. The most accomplished generals, the most disciplined and daring infantry the world has ever known, the best equipped and most extensive navy, royal and mercantile, of the age, were at the absolute command of the sovereign. Such was Spain."

Such is not Spain to-day. A quite recent writer, speaking of Spain before the war, said, that although Spain in extent holds the sixth place in the European states, "it really now subsists merely by the sufferance of stronger nations." Thus has that nation, which three centuries ago dominated the world, lost both its position and its energy.

Without attempting to sketch chronologically, either this rise or this decline, let us rather direct our efforts to an inquiry into the causes of both the one and the other.

In attempting to explain the greatness of Spain we must give first place to the vigor of the Spanish race. The great Spaniard was a mighty compound. He had the blood of Rome mingled with the awful torrent that gave birth to the soulless Goths and Vandals. In him also flowed the hot blood of the Moors. He was both sturdy and fiery; he had the fervor of the South with the tenacity of the North; the pride of the Roman with the passion of the Moor. The Spanish race was emphatically a rich race.

And then we must remember that this race had been forged in war. Century after century, from the earliest times, they had lived with their arms in their hands. First came the long war between the Arian Vandals, and the Trinitarian natives; then the seven-hundred-year war with the followers of Mahomed. The whole mission of life to them was to fight.

Naturally there was developed in the people at large the most complete unification and subjection. Individualism gave place almost entirely to the common weal, and the spectacle was presented of a nation with no political questions. Maccaulay maintains that human nature is such that aggregations of men will always show the two principles of radicalism and conservatism, and that two parties will exist in consequence, one composed of those who are ever looking to a brighter future, the other of those who are ever seeking to restore a delightful past; but no such phenomena appear in the ascending period of Spain's history. The whole nation moved as an organized army, steadily forward, until its zenith was reached. This solidity was a marked element of its strength.

Mr. Buckle recognizes this, and accounts for the harmonious movements of the nation by the influence of two leading principles, which he is pleased to call superstition and loyalty.

The Arab invasion had pressed upon the Christians with such force that it was only by the strictest discipline that the latter had managed to survive. To secure such discipline, and at the same time supply the people with the steady enthusiasm necessary to support a war from century to century, all the terrors and all the glories that could be derived from religion were employed. The church and the state, the prince and the priest, became as one, and loyalty and religion, devotion to the standard and to the cross, were but different names for the same principles and actions. Hence Spain emerged to greatness without the least dream of liberty of either person, conscience or thought. Her rallying cry was: For the Prince and the Church; not, For God and Liberty. She went up to greatness the most loyal and the most religious of nations; but Liberty, Justice and Truth were not upon her banners.

Look over the territory settled and conquered by her, and what do we see? Columbus, sailing under Spain, names the first land he discovers San Salvador; the first settlement made in this country is St. Augustine; the second, Sante Fe. Look down over the southern half of our continent and such names as Espirito Santo, Corpus Christi, San Diego, San Juan, San Jose, San Domingo attest the religious zeal of the conquerors. They were missionaries of the Cross, robbing the people of their gold and paying them off with religion.

Steadfast in the faith and sturdy in her loyalty, Spain resisted all innovations with respect to her religious beliefs, and all insurrections against her government. Her Alva and her Torquemada but illustrated how strong was her conservatism, while her Isabella and her Philip II show how grand and comprehensive and how persistent was her aggressiveness, under the idea of spreading and upholding the true faith. She not only meant to hold all she had of wealth and power, but she aspired to universal dominion; already chief, she desired to be sole, and this in the interest and name of the Holy Church.

The Reformation did not disturb Spain; it was crushed out within twenty years. The spirit of liberty that had been growing in England since Bosworth's Field, and that was manifesting itself in Germany and the Netherlands, and that had begun to quiver even in France, did not dare stir itself in Spain. Spain was united, or rather, was solidity itself, and this solidity was both its strength and its death. England was not so united, and England went steadily onward and upward; but Spain's unity destroyed her, because it practically destroyed individualism and presented the strange paradox of a strong nation of weak men.

As a machine Spain in the sixteenth century was a marvel of power; as an aggregation of thinking men, it was even then contemptible. Ferdinand, Charles V and Philip II were able and illustrious rulers, and they appeared at a time when their several characters could tell on the immediate fortunes of Spain. They were warriors, and the nation was entirely warlike. During this period the Spaniard overran the earth, not that he might till the soil, but that he might rob the man who did. With one hand he was raking in the gold and silver of Mexico and Peru; with the other confiscating the profits of the trade and manufactures of the Low Countries—and all in the name of the Great God and Saints!

How was Spain overthrown? The answer is a short one. Spain, under Philip II staked her all upon a religious war against the awakening age. She met the Reformation within her own borders and extinguished it; but thought had broken loose from its chains and was abroad in the earth. England had turned Protestant, and Elizabeth was on the throne; Denmark, Norway and Sweden, indeed all countries except Spain and Italy had heard the echoes from Luther's trumpet blast. Italy furnished the religion, and Spain the powder, in this unequal fight between the Old and the New. Spain was not merely the representative of the old, she WAS the old, and she armed her whole strength in its behalf.

Here was a religion separated from all moral principle and devoid of all softening sentiment—its most appropriate formula being, death to all heretics. Death—not to tyrants, not to oppressors, not to robbers and men-stealers—but death to *heretics*. It was this that equipped her Armada.

The people were too loyal and too pious to THINK, and so were hurled in a solid mass against the armed thought of the coming age, and a mighty nation crumbled as in a day. With the destruction of her Armada her warlike ascendancy passed and she had nothing to put in its place. She had not tillers of the soil, mechanics or skilled merchants. Business was taking the place of war all over the world, but Spain knew only religion and war, hence worsted in her only field, she was doomed.

From the days of Philip II her decline was rapid. Her territory slipped from her as rapidly as it had been acquired. Her great domains on our soil are now the seat of thriving communities of English-speaking people. The whole continent of South America has thrown off her yoke, though still retaining her language, and our troops now embarked from Port Tampa are destined to wrest from her the two only remaining colonies subject to her sway in the Western World.—Cuba and Porto Rico. With all her losses hitherto, Spain has not learned wisdom. Antagonistic to truth and liberty, she seems to sit in the shadow of death, hugging the delusions that have betrayed her, while all other people of earth are pressing onward toward light and liberty.

The struggle in Cuba had been going on for years, and in that colony of less than two millions of inhabitants, many of whom were Spaniards, there was now an army four times as large as the standing army of the United States. Against this army and against the Government of Spain a revolt had been carried on previous to the present outbreak for a period of ten years, and which had been settled by concessions on the part of the home government. The present revolt was of two years' standing when our government decided to interfere. The Cubans had maintained disorder, if they had not carried on war; and they had declined to be pacified. In their army they experienced no color difficulties. Gomez, Maceo and Quintin Banderas were generals honored and loved, Maceo especially coming to be the hero and idol of the insurgents of all classes. And it can truthfully be said that no man in either the Cuban or Spanish army, in all the Cuban struggle previous to our intervention, has earned a loftier fame as patriot, soldier and man of noble mould than ANTONIO MACEO.

Cuba, by far the most advanced of all the West Indian colonies; Cuba, essentially Spanish, was destined to be the battle ground between our troops and the veterans of Spain. The question to be settled was that of Spain's sovereignty. Spain's right to rule over the colonies of Cuba and Porto Rico was disputed by the United States, and this question, and this alone, is to be settled by force of arms. Further than this, the issue does not go. The dictum of America is: Spain shall not rule. The questions of Annexation, Expansion and Imperialism were not before us as we launched our forces to drive Spain out of the West Indies. The Cuban flag was closely associated with our own standard popularly, and "Cuba Libre" was a wide-spread sentiment in June, 1898. "We are ready to help the Cubans gain their liberty" was the honest expression of thousands who felt they were going forward in a war for others.

CHAPTER V.

PASSAGE, LANDING, AND FIRST BATTLE IN CUBA.

The Tenth Cavalry at Guasimas—The "Rescue of the Rough Riders"—Was There an Ambush?—Notes.

"The passage to Santiago was generally smooth and uneventful," says General Shafter in his official report. But when the fact is called to mind that the men had been on board a week before sailing, and were a week more on the passage, and that "the conveniences on many of the transports in the nature of sleeping accommodations, space for exercise, closet accommodations, etc., were not all that could have been desired," and that the opinion was general throughout the army that the travel ration was faulty, it cannot be doubted that the trip was a sore trial to the enlisted men at least. The monotonous days passed in the harbor at Port Tampa, while waiting for orders to sail, were unusually trying to the men. They were relieved somewhat by bathing, swimming, gaming and chatting on the coming events. A soldier who was in one of the colored regiments describes the inside life of one of the transports as follows: "After some miles of railroad travel and much hustling we were put on board the transport. I say *on board*, but it is simply because we cannot use the terms *under board*. We were huddled together below two other regiments and under the water line, in the dirtiest, closest, most sickening place imaginable. For about fifteen days we were on the water in this dirty hole, but being soldiers we were compelled to accept this without a murmur. We ate corn beef and canned tomatoes with our hard bread until we were anything but half way pleased. In the fifth or sixth day out to sea the water furnished us became muddy or dirty and well flavored with salt, and remained so during the rest of the journey. Then, the ship's cooks, knowing well our condition made it convenient to themselves to sell us a glass of clean ice water and a small piece of bread and tainted meat for the sum of seventy-five cents, or one dollar, as the case might be."

A passage from Port Tampa, around the eastern end of Cuba, through the Windward Passage, even in June, is ordinarily pleasant. On the deck of a clean steamer, protected from the sun's rays by a friendly awning, it may be put down as nearly an ideal pleasure trip; but crowded into freight ships as these men were, many of them clad in thick and uncomfortable clothing, reduced to the uninviting travel ration, compelled to spend most of the time below decks, occupied with thoughts of home and friends, and beset with forebodings of coming events, it was very far from being to them a pastime. Of the thousands who are going to Cuba to magnify the American flag, not all will return. Occasionally the gay music of the bands would relieve the dull routine and cause the spirits to rise under the effects of some enlivening waltz or stirring patriotic air; or entering a school of flying fish the men would be entertained to see these broad-finned creatures dart from the waves like arrows from the bow, and after a graceful flight of perhaps near two hundred yards drop again into the sea; but taken altogether it was a voyage that furnishes little for the historian.

The transports were so arranged as to present an interesting and picturesque spectacle as they departed from our shores on their ocean march. Forming in three columns, with a distance of about 1,000 yards between the columns, and the vessels in the columns being distanced from one another about 400 yards, the fleet was convoyed from Port Tampa by small naval vessels until it reached a point between the Dry Tortugas and Key West. Here it was met by the noble battleship Indiana and nine other war vessels, thus making a convoy altogether of fifteen fighting craft. Transports and convoy now made an armada of more than forty ships, armed and manned by the audacious modern republic whose flag waved from every masthead. Thus spreading out over miles of smooth sea, moving quietly along by steam, carrying in its arms the flower of the American army, every man of which was an athlete, this fleet announced to the world the grim purpose of a nation aroused.

The weather from the time of leaving Port Tampa continued fine until the fleet entered the passage between the western coast of Hayti and the eastern end of Cuba, known as the Windward Passage, when the breeze freshened and a rough sea began, continuing more or less up to the time of landing. Rounding this eastern coast of Cuba the fleet headed its course westerly and on the morning of the 20th was able to determine its position as being off Guantanamo Bay, about fifty miles east of Santiago. Here, eight days before, the first battle on Cuban soil, in which four American marines were killed, had been fought. About noon on the same day, the fleet came to a halt off Santiago harbor, or a little to the west of the entrance to it, and Admiral Sampson came on board. He and General Shafter soon after went ashore to consult the Cuban General, Garcia, who was known to be in that vicinity with about 4,000 well armed troops.

The voyage over, and the men having been crowded together

on shipboard for nearly two weeks, it was now expedient to get them on shore as soon as possible. But it was necessary to find out beforehand what defences were along the coast, and what forces of the enemy were likely to be encountered in landing. The fleet lay off from the shore about a mile, and it was no small undertaking to convey the 17,000 men on board with all their arms and equipments to the shore in small boats over a rough sea, especially should the landing be disputed. It was to arrange for the landing and also to map out a general plan of campaign that the three great leaders, Shafter, Sampson and Garcia met at Aserradores on the afternoon of June 20th as the American fleet stood guard over the harbor of Santiago.

General Garcia was already aware of the coming of the fleet, having received a message from Major-General Miles two weeks previous. The letter of General Miles ran as follows:

Headquarters of the Army,
In the Field, Tampa, Fla., June 2, 1898.

Dear General:—I am very glad to have received your officers, General Enrique Collazo and Lieut.-Col. Carlos Hernandez, the latter of whom returns to-night with our best wishes for your success.

It would be a very great assistance if you could have as large a force as possible in the vicinity of the harbor of Santiago de Cuba, and communicate any information by signals which Colonel Hernandez will explain to you either to our navy or to our army on its arrival, which we hope will be before many days.

It would also assist us very much if you could drive in and harass any Spanish troops near or in Santiago de Cuba, threatening or attacking them at all points, and preventing, by every means, any possible re-enforcement coming to that garrison. While this is being done, and before the arrival of our army, if you can seize and hold any commanding position to the east or west of Santiago de Cuba, or both, that would be advantageous for the use of our artillery, it will be exceedingly gratifying to us."

To this General Garcia replied that he would "take measures at once to carry out your (Miles') recommendation, but concentration of forces will require some time. Roads bad and Cubans scattered. Will march without delay." Admiral Sampson also cabled the Secretary of the Navy that Garcia "regards his (Miles') wishes and suggestions as orders, and immediately will take measures to concentrate forces at the points indicated, but he is unable to do so as early as desired on account of his expedition at Banes Port, Cuba, but will march without delay. All of his subordinates are ordered to assist to disembark the United States troops and to place themselves under orders." It was in compliance with these requests that General Garcia had the five thousand troops so near Santiago at the time he welcomed Shafter and Sampson to his camp, as mentioned above, and there is every necessary evidence that these Cuban troops took part in the fight about Santiago. Says General Miles of Garcia: "He had troops in the rear as well as on both sides of the garrison at Santiago before the arrival of our troops."

It was agreed that the force of five hundred men under General Castillo, posted near Daiquiri, should be increased to 1,000, and should be prepared to make an attack upon the rear of the Spanish garrison at Daiquiri on the morning of the 22nd, at which time the debarkation would begin. General Rabi with about 500 men was also to attack Cabanas at the same time, in the same manner, the transports and war vessels so manœuvring as to give the impression that a landing was to be made at that place. While these attacks in the rear were distracting the garrisons, the navy, by order of Admiral Sampson, was to start up a vigorous bombardment of all the villages along the coast, thus clearing the shore for the landing of the army. Thus did the conference unite the hands of Americans and Cubans in the fight against Spain on Cuban soil, and each was pledged to the other by the expressions of good will. Having accomplished its work the important conference closed, Admiral Sampson and General Shafter to return to their ships, and General Garcia to carry out the part of the work assigned to him, which he did with fidelity and success.*

According to orders published on the 20th, General Lawton's Division, known as the Second Division, Fifth Army Corps, was to disembark first. This Division contained the three following Brigades: The First, General Ludlow's, composed of the Eighth and Twenty-second Infantry (regulars) and the Second Massachusetts Volunteer Infantry; the Second Brigade, General Miles', composed of the Fourth and Twenty-fifth Infantry (regulars); the Third Brigade, General Chaffee's, containing the Seventh, Twelfth and Seventeenth Infantry (regulars). Next to follow was General Bates' Brigade, which was to act as reserve to Lawton's Division. This Brigade consisted of the Third and Twentieth Infantry (regulars) and one squadron of the Second Cavalry, the only mounted troops in Shafter's army. The cavalry, however, were not to disembark with the Brigade, but were to be the last troops to leave the transports. After Bates' Brigade, was to follow Wheeler's Dismounted Cavalry Division, containing the two following Brigades: The First, composed of the Third, Sixth and Ninth Cavalry (regulars); the Second, composed of the First and Tenth Cavalry (regulars) and the First Volunteer Cavalry (Rough Riders). To follow the Cavalry Division was to come the First Division, General Kent's, containing the following troops: The First Brigade, General

*See Note A at the end of this chapter.

Hawkins', consisting of the Sixth and Sixteenth Infantry (regulars) and the Seventy-first New York Volunteer Infantry; the Second Brigade, General Pearson's, consisting of the Second, Tenth and Twenty-first Infantry (regulars); the Third Brigade, Colonel Wikoff's, made up of the Ninth, Thirteenth and Twenty-fourth Infantry (regulars). Then, lastly, was to depart the squadron of mounted cavalry.

Thus prepared, both on board the ships and on shore, the morning of the 22nd dawned to witness the beginning of mighty operations. The war vessels, drawn up in proper order, early began to hurl shot and shell upon the towns, forts, blockhouses and clumps of trees that could be discovered along the shore. The cannonading lasted between two and three hours and was furious throughout. Meanwhile General Lawton's Division began the work of going ashore. The sea was rough and the passage to the shore was made in small boats furnished from the transports and from the naval vessels, towed by steam launches belonging to the navy. The larger of the boats were capable of carrying ten or twelve men each, while the smaller ones could carry but six or seven. During the passage to the shore several of the men who had escaped thus far, were taken with seasickness, greatly to the amusement of their more hardy companions. The landing was made at a pier which had been used formerly as a railroad pier, but was now abandoned and somewhat dilapidated. To get from the boats to the pier in this rough sea was the most perilous part of the whole trip from Tampa to Cuba. As the boats would rise on the waves almost level with the landing place. it was necessary to leap quickly from the boat to the shore. In this way two cavalrymen of the Tenth lost their lives, falling into the sea with their equipments on and sinking before help could reach them. Some of the boats were rowed ashore and made a landing on the beach some distance from the pier. By this method some men of the Twenty-fifth tried to be the first to land, but failed, that regiment landing, however, in the first body of troops to go ashore, and being the second in order, in the invasion of the island. By night of the 22nd more than one-third of the troops were on shore, and by the evening of the 24th the whole army was disembarked according to the program announced at the beginning, the squadron of cavalry coming in at the close of the march to the shore.

The only national movement on our part deserving to be brought into comparison with the expedition against the Spanish power in Cuba, is that of fifty years earlier, when General Scott sailed at the head of the army of invasion against Mexico. Some of the occurrences of that expedition, especially connected with its landing, should be carefully studied, and if the reports which have reached the public concerning it are truthful, we would do well to consider how far the methods then in use could be applied now. Scribner's recent history, published just before the outbreak of the Spanish War, tells the story of that expedition, so far as it tells it at all, in the following sentence: "On the 7th of March, the fleet with Scott's army came to anchor a few miles south of Vera Cruz, and two days later he landed his whole force—nearly twelve thousand men— by means of surf-boats." A writer in a recent number of *The Army and Navy Journal* says General Worth's Division of 4,500 men were landed in one hour, and the whole force was landed in six hours, without accident or confusion. In the prosecution of that unholy war, which lasted about a year, nearly three thousand men were lost in battle and about as many more by disease, peace being finally made by the cession of territory on the part of Mexico, the United States paying in return much more than the territory was worth. The twenty millions paid to Texas probably in great part went into the coffers of the patriots who occupied that region, some of whom had not been known as desirable citizens in the parts from which they came, and had manifested their patriotism by leaving their country for their country's good. The fifteen millions handed over to Mexico looks like a contribution to a conscience fund, and an atonement offered for an assault without provocation. The country gained Arizona, New Mexico, California and finally Texas, but it lost six thousand good men, the cost of the war, and all told, in negotiations, about thirty million dollars, besides. However, it is not always profitable to look up the harvests of war. There are always two—the harvest of gain, and the harvest of loss. Death and debt are reapers, as well as are honor and extent of territory.

The feelings of the six thousand American troops who landed on Cuban soil on June 22nd, 1898, may well be imagined. Although they felt the effects of the confinement to which they had been subjected while on shipboard, there was very little sickness among them. Again possessed of the free use of their limbs they swarmed the beach and open space near the landing, making themselves at home, and confronting the difficulties and perils that lay before them with a courage born of national pride. Before them were the mountains with their almost impassable roads, the jungles filled with poisonous plants and the terrible prickly underbrush and pointed grass, in which skulked the land crab and various reptiles whose bite or sting was dangerous; twenty miles of this inhospitable country lay between them and Santiago, their true objective. And somewhere on the road to that city they knew they were destined to meet a well-trained foe, skilled in all the arts of modern warfare, who would contest their advance. The prospect, however, did not unnerve them, although they could well conjecture that all who landed would not re-embark. Some in that six thousand were destined never again to set foot on shipboard. Out of the Twenty-fifth Infantry and the Tenth Cavalry men were to fall both before Spanish bullets and disease ere these organizations should assemble to return to their native shores. These thoughts did not prevent the men from taking advantage of what nature had to offer them.

"We landed in rowboats, amid, and after the cessation of the bombardment of the little hamlet and coast by the men-of-war and battle-ships," writes a brave soldier of the Twenty-fifth Infantry, and adds immediately: "We then helped ourselves to cocoanuts which we found in abundance near the landing." Ordinarily this statement, so trivial and apparently unimportant, would not merit repetition, but in its connection here it is significant as showing the immediate tendency of the men to resort to the fruits of the country, despite all warnings to the contrary. The two weeks' experience on board the transports had made the finding of cocoanuts an event to be noted, and the dry pulp and strongly flavored milk of this tropical fruit became extremely grateful to the palate, even if not altogether safe for the stomach. If ripe, however, the cocoanut could scarcely be more ungenial to many, than the raw, canned tomatoes upon which they had in part subsisted during the voyage. It is to be added that this report of the finding of the cocoanuts is not the report of an old soldier, but of a young and intelligent, first enlistment man.

Lawton's Division soon after landing, was ordered to move forward in the direction of Santiago, on the road leading past Siboney. A staff officer, writing of that movement, says: "General Lawton, with his Division, in obedience to this order, pushed forward from Daiquiri about five miles, when night overtook him and he bivouacked on the road." An old soldier of the Twenty-fifth, writing me from the hospital in Tampa, Florida, July 22nd, says of the same event: "After the regiment landed we marched about four and a half miles through the mountains; then we made camp." The old soldier says nothing of cocoanuts, but makes his statement with as much accuracy as possible, and with no waste of words. The novice describing the same thing says: "A short distance ahead (from the shore) we bivouacked for the night. We were soon lying in dreamland, so far from friends and home, indeed, on a distant, distant shore." These two extracts show at once the difference between the soldier produced by years of trial and training on our plains, and the soldier who but yesterday was a civilian. With the one the march is a short distance; with the other it is about four and a half miles; one reports that they "made camp," the other talks of dreamland, friends, home and distant shore; one expresses his feelings, the other shows control of feeling and reserve in expression.

That first night on Cuban soil, the night following June 22nd, was one without events, but one of great concern to the commanders on shore and on the fleet. The work of disembarking had gone on successfully, and already about six thousand men were on shore. Nearly the whole of Lawton's Division, with Bates' independent brigade, were bivouacked, as we have seen, about five miles from Daiquiri, exactly where the railroad crosses the wagon road leading to Siboney. General Wheeler's troops—one brigade—were encamped on the open ground near the landing, the remainder of his division being still on the transports. The Twenty-fifth Infantry was with Lawton; the Tenth Cavalry was ashore with Wheeler's troops. A detachment of the Twenty-fifth was put on outpost duty on that night of their landing, and five miles within Cuban territory they tramped their solitary beats, establishing and guarding the majestic authority of the United States.

Lawton's orders were to seize and hold the town of Siboney at which place Kent's Division, containing the Twenty-fourth, was to land. It was then intended that the whole army should advance as rapidly as would be consistent with supplying the men with rations toward Santiago. Siboney was to be the base of supplies, and from this point ammunition and food were to be conveyed to the front by wagons and pack trains. General Shafter also intended that Lawton with his division should lead the advance upon Santiago, but circumstances beyond his control brought about a different result. On the morning of the 23rd Lawton's division was in motion early, and before half-past ten o'clock he was able to report that the Spaniards had evacuated Siboney and were in full retreat, pursued by a body of Cubans under direction of General Castillo; that the town was in his hands, and he had also captured one locomotive and nearly one hundred cars loaded with coal.

General Young's brigade of General Wheeler's cavalry division, got on shore on the afternoon of the 23rd and after landing received verbal orders to move out with three days' rations "to a good camping place between Juraguacito and Siboney, on the road leading to Santiago de Cuba." In obedience to these orders, at 4.30 in the afternoon Young with the Rough Riders and a squadron from each of the First and Tenth Regular Cavalry moved from the bivouack near the landing and arrived at Siboney at about 7 o'clock. When General Young arrived at Siboney he had with him the Rough Riders, the other troops having been delayed by the crowded condition of the trail and the difficulty of following after nightfall. Although these troops are always spoken of as cavalry, the reader must not forget that they were dismounted and in marching and fighting were the same as infantry.

General Young on arriving at Siboney reported to General Wheeler, who had preceded him to the same place. The statements of the several commanders here appear somewhat conflicting, although not inexplicable. General Lawton says: "Yesterday afternoon, late, General Wheeler and staff arrived and established his headquarters within the limits of my command. Saw him after dark. Late last night Colonel Wood's regiment of dismounted cavalry (Rough Riders) passed through my camp at Division Headquarters, and later General Young, with some of the dismounted Cavalry, and early this morning others of the dismounted cavalry." Wheeler says that "in obedience to instructions from the Major-General Commanding," given to him in person, he proceeded, on June 23rd, to Siboney, but does not say at what hour. He says he "rode out to the front and found that the enemy had halted and established themselves at a point about three miles from Siboney." He then informs us that "at 8 o'clock on that evening of the 23rd General Young reached Siboney with eight troops of Colonel Wood's regiment (A, B, D, E, F, G, K and L), 500 strong; Troops A, B, G and K, of the First Cavalry, in all 244, and Troops A, B, E and I, of the Tenth Cavalry, in all 220 men, making a total force of 964 men, which included nearly all of my command which had disembarked. These troops had marched from Daiquiri, 11 miles. With the assistance of General Castillo a rough map of the country was prepared and the position of the enemy fully explained, and I determined to make an attack." Lieutenant Miley says that the whole brigade of Wheeler's troops arrived in Siboney about dark and were occupying the same ground as General Lawton ("In Cuba With Shafter," p. 76.) General Young says that after reporting to General Wheeler he "asked and obtained from General Wheeler authority to make a reconnoisance in force" for the purpose of obtaining "positive information * * * as to the position and movements of the enemy in front."

The distance from Daiquiri to Siboney was but eleven miles, and as the troops left the former place at 4.30 it is probable that they were all bivouacked near Siboney before 9 o'clock, as they were all together, according to General Wheeler's report, at 5.45 on the morning of the 24th. General Young having discovered that there were two roads or trails leading from Siboney northward toward the town of Sevilla determined to make his reconnoisance by both these trails. He directed Colonel Wood to move by the western trail and to keep a careful lookout and to attack any Spaniards he might encounter, being careful to join his right in the event of an engagement, with the left of the column advancing by the eastern trail. Colonel Wood's column was the left column and was composed of the Rough Riders only. The column marching by the eastern trail was composed of the First and Tenth Cavalry (regulars) and was under the command of General Young. It was the intention of General Young by this column to gain the enemy's left, and thus attack in front and left. As early as 7.20 a. m. Captain Mills discovered the enemy exactly as had been described by General Castillo. When this was done word was sent to Colonel Wood, who was making his way to the front over a more difficult route than the one by which General Young's column had marched. A delay was therefore made on the part of General Young in order that the attack should begin on both flanks at the same time. During this delay General Wheeler arrived and was informed of the plans and dispositions for the attack, and after examining the position gave his approval of what had been done, whereupon General Young ordered the attack. General Wheeler in speaking of the same event says: "General Young and myself examined the position of the enemy. The lines were deployed and I directed him to open fire with the Hotckiss gun. The enemy replied and the firing immediately became general." There can be no question as to the planning of this fight nor as to the direction of the American force in the fight so far as any general direction was possible. Colonel Wood directed one column and General Young another, while the plan of the attack undoubtedly originated with General Young. General Wheeler conveys as much when he says: "General Young deserves special commendation for his cool deliberate and skillful management." General Young, if only the commander of the right column consisting of two squadrons of regular cavalry, had not as large a command, nor as difficult and important a one as had Colonel Wood, and hence is not deserving of special commendation except upon the general ground that he had supervision over the whole battle. This position is taken by General Shafter in his report, who though admitting the presence of the Division Commander, credits the battle to General Young, the commander of the brigade. The reconnoissance in force for which Young had obtained authority from General Wheeler on the night of the 23rd had developed into a battle, and the plan had evolved itself from the facts discovered. This plan General Wheeler approved, but in no such way as to take the credit from its originator; and it is doubtless with reference both to the plan and the execution that he bestows on General Young the mead of praise. This statement of fact does not in the least detract from either the importance or the praiseworthiness of the part played by Colonel Wood. Both he and the officers and men commanded by him received both from General Young and from the division commander the most generous praise. The advance of Wood's column was made with great difficulty owing to the nature of the ground, and according to General Young's belief, he was in the rear when at 7.20 in the morning Captain Mills discovered the enemy, and a Cuban guide was dispatched to warn Wood, and a delay made to allow time for him to come up. Colonel Wood, on the other hand, claims to have discovered the enemy at 7.10 and to have begun action almost immediately, so that it turned out as Young had planned, and "the attack of both wings was simultaneous." The Spaniards were posted on a range of high hills in the form of a "V," the opening being toward Siboney, from which direction the attack came.

From Colonel Wood's report it appears that soon after the firing began he found it necessary to deploy five troops to the right, and left, leaving three troops in reserve. The enemy's lines being still beyond his, both on the right and on the left, he hastily deployed two more troops, which made the lines now about equal in length. The firing was now "exceedingly heavy," and much of it at short range, but on account of the thick underbrush it was not very effective; "compara tively few of our men were injured." Captain Capron at this time received his mortal wound and the firing became so terrific that the last remaining troop of the reserve was absorbed by the firing line, and the whole regiment ordered to advance very slowly. The Spanish line yielded and the advance soon showed that in falling back the enemy had taken a new position, about three hundred yards in front of the advancing regiment. Their lines extended from 800 to 1,000 yards, and the firing from their front was "exceedingly heavy" and effective. A "good many men" were wounded, "and several officers," says Colonel

Wood's report. Still the advance was kept up, and the Spanish line was steadily forced back. "We now began," says Colonel Wood, "to get a heavy fire from a ridge on our right, which enfiladed our line." The reader can at once see that although the Rough Riders were advancing heroically, they were now in a very serious situation, with an exceedingly heavy and effective fire striking them in front, and a heavy, enfilading fire raking them from the right. Their whole strength was on the line, and these two fires must have reduced their effectiveness with great rapidity had it kept up, the Spaniards having their range and firing by well-directed volleys. It was for the regiment a moment of the utmost peril. Had they been alone they must have perished.

It was from this perilous situation of Colonel Wood's command that one of the most popular stories of the war originated, a story that contained some truth, but which was often told in such a way as to cause irritation, and in some instances it was so exaggerated or mutilated in the telling as to be simply ridiculous. On the day after the battle the story was told in Lawton's camp according to the testimony of an intelligent soldier of the Twenty-fifth Infantry. His words are: "The next day about noon we heard that the Tenth Cavalry had met the enemy and that the Tenth Cavalry had rescued the Rough Riders. We congratulated ourselves that although not of the same branch of service, we were of the same color, and that to the eye of the enemy we, troopers and footmen, all looked alike." According to artists and cheap newspaper stories this rescuing occurred again and again. A picture is extensively advertized as "an actual and authoritative presentation of this regiment (the Tenth Cavalry) as it participated in that great struggle, and their heroic rescue of the Rough Riders on that memorable *July* day." This especial rescuing took place on *San Juan Hill*. The editor of a religious paper declares that it was the *Twenty-fifth Infantry* that rescued the Rough Riders and that it was done at *El Caney!**

Before we go any farther let us see just what the Tenth Cavalry did do in this fight. That their action was highly meritorious admits of no doubt, and the laurels they won were never allowed to fade during the whole campaign. General Wheeler speaks of them with the First Cavalry. He says: "I was immediately with the troops of the First and Tenth Regular Cavalry, dismounted, and personally noticed their brave and good conduct." There were four troops of the Tenth engaged, composing the First Squadron of that regiment, under command of Major Norval. Troop A was commanded by Captain W. H. Beck, who was specially commended by General Wheeler for good conduct. Second Lieutenant F. R. McCoy was Captain Beck's assistant. This troop moved over to the left, receiving the fire of the enemy, but making no response, the distance being too great for effective carbine firing.

---

*THE TWENTY-FIFTH AT EL-CANEY.

American valor never shone with greater luster than when the Twenty-fifth Infantry swept up the sizzling hill of El-Caney to the rescue of the rough riders. Two other regiments came into view of the rough riders. But the bullets were flying like driving hail; the enemy were in trees and ambushes with smokeless powder, and the rough riders were biting the dust and were threatened with annihilation.

A rough rider described the feelings of his brigade when they saw the other regiments appear and retreat. Finally this rough rider, a Southerner, heard a well-known yell. And out of the distance moved a regiment as if on dress parade, faces set like steel, keeping step like a machine, their comrades falling here, there, everywhere, moving into the storm of invisible death without one faltering step, passing the rough riders, conquering up the hill, and never stopping until with the rough riders El-Caney was won. This was the Twenty-fifth Regiment (colored), United States Infantry, now quartered at Fort Logan, Denver. We have asked the chaplain, T. G. Steward, to recite the events at El-Caney. His modesty confines him to the barest recital of "semi-official" records. But the charge of the Twenty-fifth is deserving of comparison with that of "the Light Brigade" in the Crimean War, or of Custer at the massacre of the Big Horn.
(Editorial in religious paper.)

This troop reached Colonel Wood's right and made the line continuous so that there was now a force in front of that ridge where the Spaniards were securely entrenched and from which they were pouring their enfilading fire upon Colonel Wood's line. Troop A, although coming into the line, did not fire. Their presence, however, gave the Rough Riders the assurance that their flank was saved. Troop E was commanded by Captain C. G. Ayres with Second Lieutenant George Vidmar. This troop was placed by General Young in support of Captain Watson's two Hotchkiss guns, and also of the troops in their front. The troop was under fire one hour and a quarter, during which they were in plain view of the Spaniards, who also had their exact range. One man was killed and one wounded. Their courage, coolness and discipline in this trying hour and a quarter were of the very highest order. The troop commander says: "Their coolness and fine discipline were superb." This troop did not fire a shot. Thus one-half of the squadron moved to its positions and held them without being able to do any damage to the enemy, as they were carrying out to the letter their instructions, which were to fire only when they could see the enemy. Troop B was commanded by Captain J. W. Watson with H. O. Willard as Second Lieutenant. A detachment of this troop was placed in charge of four Hotchkiss mountain guns. This detachment opened fire upon the enemy, using the ammunition sparingly, as they had but fifty rounds with them. Twenty-two shots were fired, apparently with effect. The remainder of the troop under Lieutenant Williard was ordered to move out to the extreme right, which would place it beyond the line of the First Cavalry, thus bringing that regiment between Troop A of the Tenth, which connected it with the Rough Riders and Troop B, which was to be on its extreme right. Lieutenant Williard's report of this movement is as follows:

"I ordered the troop forward at once, telling them to take advantage of all cover available. In the meantime the volleys from the Spanish were coming in quite frequently and striking the ground on all sides near where we were. I found it very difficult to move the men forward after having found cover, and ran back to a portion of the troop near an old brick wall, and ordered them forward at once. They then made a dash forward, and in doing so three or four men were wounded, Private Russell severely. Who the others were I do not know. We encountered a severe fire directly after this move forward; and Private Wheeler was wounded in the left leg. There was a wire fence on our right, and such thick underbrush that we were unable to get through right there, so had to follow along the fence for some distance before being able to penetrate. Finally, was able to get the greater proportion of my men through, and about this time I met Lieutenants Fleming and Miller, Tenth Cavalry, moving through the thicket at my left. I there heard the order passed on 'not to fire ahead,' as there was danger of firing into our own forces. In the meantime there was shouting from the First Cavalry in our front, 'Don't fire on us in rear.' My troop had not fired a shot to my knowledge, nor the knowledge of any non-commissioned officers in the troop. About this time I found I was unable to keep the troop deployed, as they would huddle up behind one rock or tree, so I gave all sergeants orders to move out on the extreme right and to keep in touch with those on their left. Then, with a squad of about five men, I moved to the right front, and was unfortunate enough to lose the troop, i. e., I could see nothing of them except the men with me.

"But as I had given explicit instructions to my sergeant, in case I was lost from them, to continue to advance until halted by some one in authority, I moved ahead myself, hoping to find them later on. In making a rush forward three men of my squad were lost from me in some way. I still had two men with me, Privates Combs and Jackson, and in the next advance made I picked up a First Cavalry sergeant who had fallen out from exhaustion. After a terrific climb up the ridge in front of me, and a very regular though ineffective fire from the enemy kept up until we were about sixty yards from the summit of hill, we reached the advance line of the First United States Cavalry, under command of Captain Wainwright. I then reported to him for orders, and moved forward when he next advanced. The firing had ceased, and no more shots wer fired, to my knowledge, after this time. With the First Cavalry, Troop G, we followed along the right of the ridge and came down to the right front, encountering no opposition or fire from the enemy, but finding the enemy's breastworks in confusion, ammunition and articles of clothing scattered around; also one dead Spaniard and two Mauser rifles. At the foot of the ridge we met some of the First Volunteer Cavalry, and being utterly exhausted, I was obliged to lie down. Soon after, Captain Mills, adjutant-general of Second Brigade, Cavalry Division, came up to where I was and placed me in command of Troop K, First United States Cavalry, whose officers were wounded. I then marched them forward on the road to where General Wheeler was sitting, and received orders from Colonel Wood, First Volunteer Cavalry, to remain until further orders and make no further advance. Directly afterwards, learning the action was over, I reported back to General Young, and received orders to remain camped with the First Cavalry Squadron, where the action had closed. In the meantime, I should have stated that I found the principal part of my troop and collected them and left them under the first sergeant, when I went back to receive orders. So far as I know, and to the best of my knowledge, the men of my troop acted with the greatest bravery, advancing on an enemy who could not be seen, and subjected to a severe and heavy fire at each step, which was only rendered ineffective to a great degree by the poor marksmanship of the enemy, as many times we were in sight of them (I discovered this by observation after the engagement) while we could see nothing. We were also subjected to a severe reverse fire from the hills in our right rear, several men being wounded by this fire. Throughout the fight the men acted with exceptional coolness, in my judgment. The casualties were: Privates Russell, Braxton and Morris, severely wounded; Privates F. A. Miller, Grice, Wheeler and Gaines, slightly wounded, i. e., less severely. None killed.

Very respectfully,

HENRY O. WILLIARD.

June 24, 1898.

Troop B, Tenth Cavalry, during action near La Guasima, Second Lieutenant, Tenth United States Cavalry, Commanding.

Troop I of the Tenth Cavalry was commanded by First Lieutenant R. J. Fleming with Second Lieutenant A. M. Miller. This troop moved to the right and wedged in between B Troop and the right of the First Cavalry. Lieutenant Fleming discovered the enemy posted on the high ridge immediately in front of his troop, and also extending to his right, in front of B Troop. Moving his troop a little to the right so as to secure room to advance without coming in contact with the First Cavalry, he then directed his course straight toward the hill on which he had located the enemy. The advance was made with great caution, the men seeking cover wherever possible, and dashing across the open spaces at full run. Thus they moved until the base of the steep part of the hill was reached. This was found very difficult of ascent, not only because of the rugged steepness, but also on account of the underbrush, and the sharp-leaved grass, the cacti and Spanish bayonet, that grow on all these hillsides. Paths had to be cut through these prickly obstructions with knives and sabres. Consequently the advance up that hill, though free from peril, was very slow and trying. Twice during the advance the men obtained a view of their enemies and were permitted to fire. The instructions were rigidly adhered to: No firing only at the visible foe. Lieutenant Fleming says: "Owing to the underbrush it was impossible for me to see but a very few men at a time, but as they all arrived on the crest about the time I did, or shortly after, they certainly advanced steadily." He says: "The entire troop behaved with great coolness and obeyed every order." Farrier Sherman Harris, Wagoner John Boland and Private Elsie Jones especially distinguished themselves for coolness and gallantry. The aggressive work of the Tenth Cavalry, therefore, appears to have been done by Troops B and I, a detachment of the former troop serving the Hotchkiss gun battery. Troop I was commanded by Lieutenant Fleming and by him conducted to the front, although he admits that in their advance up the slope of the hill he could see but very few of the men at a time, and declares that their advance was certainly steady, because all arrived at the crest of the hill simultaneously or nearly so.

Lieutenant Fleming does not show that his troop of excellent men were in any sense *peculiarly* dependent upon their white officers as some have asserted. They advanced steadily, just as the regulars always do, advanced noiselessly and without any reckless firing, and reached the crest of the hill in order, although he could not see them as they were making their advance. They kept their line despite all the obstructions. Lieutenant Fleming also says that in moving to his position he passed Troop B, which then "inclined to the right, and during the remainder of the action was on my right." Troop B, therefore, went through about the same experience as Troop I, and being on the extreme right of the line may have been more directly in front of that foe which Fleming says was in his front and to the right. Why did not the officer who directed or led B Troop in its advance upon the enemy report the action of his troop as vividly and generously as did Lieutenant Fleming the men of Troop I? With not the slightest reflection upon the gallant officer, he himself has the manliness to say he was so unfortunate as to lose the troop. The troop, however, did not become demoralized, but went into action under command of its First Sergeant, *John Buck,* and remained on Lieutenant Fleming's right during the action.* It has been proven more than once that should the commissioned officers of a company or troop of colored regulars be killed or incapacitated, the non-

---

*See Note C at the end of this chapter.

commissioned officers can carry on the fight. Speaking of this same regiment it is equally true that at San Juan the officers of Troops D and G were all shot and the commands of these troops fell to their First Sergeants, the first to Sergeant William H. Given, the second to Sergeant Saint Foster, and it is generally understood that these two men were appointed Lieutenants of Volunteers because of their success in handling their troops in battle.

The entire attacking force at this end of the line, if we count only those engaged in actual firing, consisted of two troops of the Tenth Cavalry and two of the First Cavalry—four troops—while to the left the entire eight troops were on the firing line. The action of the troops of the First Cavalry was quite similar to that of the troops of the Tenth Cavalry, and equally deserving of commendation. Of them all General Young says:

"The ground over which the right column advanced was a mass of jungle growth, with wire fences, not to be seen until encountered, and precipitous heights as the ridge was approached. It was impossible for the troops to keep in touch along the front, and they could only judge of the enemy from the sound and direction of his fire. However, had it not been for this dense jungle, the attack would not have been made against an overwhelming force in such a position. Headway was so difficult that advance and support became merged and moved forward under a continuous volley firing, supplemented by that of two rapid-fire guns. Return firing by my force was only made as here and there a small clear spot gave a sight of the enemy. The fire discipline of these particular troops was almost perfect. The ammunition expended by the two squadrons engaged in an incessant advance for one hour and fifteen minutes averaged less than ten rounds per man. The fine quality of these troops is also shown by the fact that there was not a single straggler, and in not one instance was an attempt made by any soldier to fall out in the advance to assist the wounded or carry back the dead. The fighting on the left flank was equally creditable and was remarkable, and I believe unprecedented, in volunter troops so quickly raised, armed and equipped."

The five hundred men of Colonel Wood's regiment were stretched over a space of 800 to 1,000 yards, and were entirely without support or reserve, and appear to have advanced to a point where this very strong force on the right swept a good part of their line both with rifle fire and the fire of their two machine guns. Men and officers were falling under both the front and flank fire of the enemy, and had not the squadrons of the First and Tenth made their successful assault upon that ridge, which, according to General Wood's report, was "very strongly held," the situation of the Rough Riders would have been extreme. Because this successful assault was participated in by the Tenth Cavalry the story arose that the Rough Riders were rescued by that regiment. The fair statement would be: That the Regular Cavalry, consisting of a squadron of the First and a squadron of the Tenth, made their advance on the right at the precise moment to deliver the Rough Riders from a fire that threatened their annihilation. The marksmanship and coolness of the men of the Tenth have been specially commented upon and their fire was described as very effective, but the same remarks could be made of the men of the First, who fought side by side with them. It is probable that the volunteers advanced more rapidly than did the regulars, using more ammunition, and manifesting a very high degree of courage and enthusiasm as well as deliberation; but the regulars reached their objective at the proper time to turn the battle's tide. Each advancing column was worthy to be companion to the other.

General Wheeler said the fire was very hot for about an hour, and "at 8.30 sent a courier to General Lawton informing him that he was engaged with a larger force of the enemy than was anticipated, and asked that his force be sent forward on the Sevilla road as quickly as possible." ("In Cuba With

Shafter," p. 83.) General Lawton, however, with the true instinct of a soldier had already sent orders to General Chaffee to move forward with the First Brigade. The Second Brigade was also in readiness to move and the men of the Twenty-fifth were expecting to go forward to take a position on the right and if possible a little to the rear of the Spanish entrenchments in order to cut off their retreat. The rapid movements of the cavalry division, however, rendered this unnecessary, and the routing of the foe gave to the Americans an open country and cleared the field for the advance on Santiago. The first battle had been fought, and the Americans had been victorious, but not without cost. Sixteen men had been killed and fifty-two wounded. In Colonel Wood's regiment eight had been killed and thirty-four wounded; in the First Cavalry, seven killed and eight wounded; in the Tenth Cavalry, one killed and ten wounded. The percentage of losses to the whole strength of the several organizations engaged was as follows: Rough Riders, over 8 per cent.; First Cavalry, over 6 per cent.; Tenth Cavalry, 5 per cent. But if we take those on the firing line as the base the rate per cent. of losses among the regulars would be doubled, while that of the volunteers would remain the same.

The strength of the enemy in this battle is given in the Spanish official reports, according to Lieutenant Miley, at about five hundred, and their losses are put at nine killed and twenty-seven wounded. At the time of the fight it was supposed to be much larger. General Young's report places the estimates at 2,000, and adds "that it has since been learned from Spanish sources to have been 2,500. The Cuban military authorities claim the Spanish strength was 4,000." These figures are doubtless too high. The force overtaken at Las Guasimas was the same force that evacuated Siboney at the approach of Lawton and the force with which the Cubans had fought on the morning of the 23rd. It may have consisted solely of the garrison from Siboney, although it is more probable that it included also those from Daiquiri and Jutici, as it is quite certain that all these troops proceeded toward Santiago over the same road. The force at Siboney had been given by the Cubans at 600, at Daiquiri at 300, and at Jutici at 150. If these had concentrated and the figures were correct, the Spanish force at Guasimas was upwards of 1,000. If, however, it was the force from Siboney alone, it was about as the Spanish official report gives it. On this latter basis, however, the losses are out of proportion, for while the attacking party lost a little less than 7 per cent. of its entire strength in killed and wounded, the losses of the entrenched, defending party, were even a little greater, or over 7 per cent. of its strength. It is, therefore, probable that the Spanish force was greater than officially reported and included the troops from the other posts as well as those from Siboney. The engagement was classed by General Shafter as unimportant, although its effect upon our army was inspiring. It did not cut off the retreat of the Spanish force, and the men who faced our army at Guasimas met them again in the trenches before Santiago. General Shafter desired to advance with his whole force, and cautioned strongly against any further forward movement until the troops were well in hand. The two battles between the Cubans and Spaniards, fought on the 23rd, in which the Cubans had sixteen men wounded and two killed, were engagements of some consequence, although we have no reports of them. There is no evidence that the Cubans took part in the battle of Guasimas, although they arrived on the grounds immediately after the firing ceased.

The story thus far told is, as the reader cannot fail to see, directly from official records, and the conclusions arrived at are those which result naturally from the facts as therein detailed. Not one word is quoted from any but military men—actors in the affair. We may now go briefly over the same ground, giving the views and conclusions of able civilian correspondents who followed the army to see what was done, and who were trained observers and skilled writers. How have these able war journalists told the story of Las Guasimas?

To quote from Stephen Bonsal in substance, not in words, is to contradict what General Shafter says officially in one particular, but in no such way as to discredit the General, or to weaken Bonsal. It is not a case of bringing two universal, antagonistic propositions face to face, but a case where two men of different training look upon an action from different standpoints and through different field-glasses. General Shafter says of the collision of the Rough Riders with the Spanish force: "There was no ambush as reported." As a military man, he says there was no more concealment on the part of the Spanish force than what an attacking party should expect, no more than what is usual in modern warfare, hence he does not regard it as an ambush, and does not officially take notice of any surprise or unexpected encounter on the part of his force. To do so would be to reflect, however slightly, upon the professional skill of the commander of the left column. General Shafter thus says officially in a manly way: "There was no ambush." Beyond this his duty does not call him to go, and he halts his expressions exactly at this line, maintaining in his attitude all the attributes of the true soldier, placing himself beyond criticism by thus securing from attack the character of his subordinate.

Mr. Bonsal is a writer and author, accustomed to view actions in the broader light of popular judgment, entirely free from professional bias, and having no class-feeling or obligations to serve. His pen is not official; his statements are not from the military standpoint; not influenced in any way by considerations of personal weal or woe with respect to others or himself. He says that one troop of the Rough Riders, Troop L, commanded by Captain Capron, was leading the advance of the regiment, and was in solid formation and within twenty-five yards of its scouting line when it received the enemy's fire. This troop was so far in the advance that it took the other troops of the regiment more than a half hour to get up to it. The writer speaks of the advance of that troop as having been made "in the fool-hardy formation of a solid column along a narrow trail, which brought them, in the way I have described, within point-blank range of the Spanish rifles, and within the unobstructed sweep of their machine guns." He sums up as follows: "And if it is to be ambushed when you receive the enemy's fire perhaps a quarter of an hour before it was expected, and when the troop was in a formation, and the only one in which, in view of the nature of the ground it was possible to advance quickly, then most certainly L Troop of the Rough Riders was ambushed by the Spaniards on the morning of June 24th."

Mr. Bonsal also brings into clear view the part taken in this battle by Lawton's Infantry. He shows by means of a simple map the trail by which Miles' brigade, in which was the Twenty-fifth Infantry, moved in order to flank the Spanish position, while Chaffee's brigade was hurrying forward on the Royal Road to reinforce the line in front. A letter from a soldier of the Twenty-fifth written soon after these events fully confirms Mr. Bonsal in what he says concerning the movement of Miles' brigade. The soldier says: "On the morning of the 24th the Rough Riders, Tenth and First Cavalry were to make an attack on a little place where the Spanish were fortified. The Second Brigade was to come on the right flank of these troops and a little in rear of the fortifications; but by some misunderstanding, the former troops, led by the Rough Riders, made an attack before we got our position, and the result was a great many lives lost in the First Cavalry and Rough Riders —only one in Tenth Cavalry, but many wounded. They captured the fortification." This letter by a humble soldier, written with no thought of its importance, shows how gallantly Lawton had sprung to the rescue of Wheeler's division. According to Bonsal, who says he obtained his information from Spanish officers who were present in this fight, it was the information of the approach of this brigade and of Chaffee's up the main road that caused the Spaniards to withdraw rapidly from the position. The whole force was in imminent danger of being captured. Another soldier of the Twenty-fifth wrote: "The report came that the Twenty-fifth Infantry was to cut off the Spanish retreat from a stronghold, toward Santiago." These glimpses from soldiers' letters illustrate how clearly they comprehended the work upon which they were sent, and show also how hearty and cordial was the support which the infantry at that time was hurrying forward to the advancing cavalry.

The official reports show that the strength of the Spanish position was before the right of our line. Mr. Bonsal says: "Directly in front of the Tenth Cavalry rose undoubtedly the strongest point in the Spanish position—two lines of shallow trenches, strengthened by heavy stone parapets." We must remember that so far as we can get the disposition of these troops from official records, Troop A connected the Rough Riders with the First Cavalry, and Troops I and B were on the right of the First Cavalry. Troop A did not fire a shot; the fighting, therefore, was done by Troops I and B on the extreme right of the line, and it was on their front that "undoubtedly the strongest point in the Spanish position" lay—nor should the reader forget that at this very important moment Troop B was commanded by its First Sergeant, Buck, Lieutenant Williard having by his own report been "unfortunate enough to lose the troop." This is said with no disparagement to Lieutenant Williard. It was merely one of the accidents of battle.

Says Mr. Bonsal: "The moment the advance was ordered the black troopers of the Tenth Cavalry forged ahead. They were no braver certainly than any other men in the line, but their better training enabled them to render more valuable services than the other troops engaged. They had with them and ready for action their machine guns, and shoved them right up to the front on the firing line, from where they poured very effective fire into the Spanish trenches, which not only did considerable execution, but was particualrly effective in keeping down the return fire of the Spaniards. The machine guns of the Rough Riders were mislaid, or the mules upon which they had been loaded could not be found at this juncture. It was said they had bolted. It is certain, however, that the guns were not brought into action, and consequently the Spaniards suffered less, and the Rough Riders more, in the gallant charge they made up the hill in front of them, after the Tenth Cavalry had advanced and driven the Spaniards from their position on the right."

Corporal W. F. Johnson, B Troop, was the non-commissioned officer in charge of the machine guns during the brief fight at Las Guasimas, and his action was such as to call forth from the troop commander special mention "for his efficiency and perfect coolness under fire." Here I may be pardoned for calling attention to a notion too prevalent concerning the Negro soldier in time of battle. He is too often represented as going into action singing like a zany or yelling like a demon, rather than as a man calculating the chances for life and victory. The official reports from the Black Regulars in Cuba ought to correct this notion. Every troop and company commander, who has reported upon colored soldiers in that war, speaks of the coolness of the men of his command. Captain Beck, of Troop A, Tenth Cavalry, in the Guasimas fight, says: "I will add that the enlisted men of Troop A, Tenth Cavalry, behaved well, silently and alertly obeying orders, and without becoming excited when the fire of the enemy reached them." The yell, in the charge of the regulars, is a part of the action, and is no more peculiar to Negro troops than to the whites, only as they may differ in the general timbre of voice. Black American soldiers when not on duty may sing more than white troops, but in quite a long experience among them I have not found the difference so very noticeable. In all garrisons one will find some men more musically inclined than others; some who love to sing and some who do not; some who have voices adapted to the production of musical tones, and some who have not, and it is doubtless owing to these constitutional differences that we find differences in habits and expressions.

Lieutenant Miley, of General Shafter's staff, in his description of the departure of General Shafter from General Garcia's tent, gives us a glimpse of the character of the men that composed the Cuban army in that vicinity.

"While the interview was going on, the troops were being assembled to do honor to the General on his departure. Several companies were drawn up in front of the tent to present arms as he came out, and a regiment escorted him to the beach down the winding path, which was now lined on both sides by Cuban soldiers standing about a yard apart and presenting arms. The scene made a strong impression on all in the party, there seemed to be such an earnestness and fixedness of purpose displayed that all felt these soldiers to be a power. About fifty per cent. were blacks, and the rest mulattoes, with a small number of whites. They were very poorly clad, many without shirts or shoes, but every man had his gun and a belt full of ammunition."

B.

EXTRACT FROM A LETTER FROM A SOLDIER OF THE 10TH CAVALRY, TROOP B, CONCERNING THE BATTLE OF LAS GUASIMAS:

" . . . The platoon which escaped this ditch got on the right of the 1st Cavalry on the firing line, and pushed steadily forward under First Sergeant Buck, being then in two squads—one under Sergeant Thompson. On account of the nature of the ground and other natural obstacles, there were men not connected with any squads, but who advanced with the line.

Both squads fired by volley and at will, at the command of the sergeants named; and their shots reached the enemy and were effective, as it is generally believed.

Private W. M. Bunn, of Sergeant Thompson's squad, is reported to have shot a sharpshooter from a tree just in front of the enemy's work. Private Wheeler was shot twice in the advance. Sergeant Thompson's squad was once stopped from firing by General Wheeler's adjutant-general for fear of hitting the Rough Riders.

It seems that two distinct battles were fought that day. Colonel Wood's command struck the enemy at about the same time, or probably a litle before, ours did, and all unknown to the men in our ranks; and got themselves into a pretty tight squeeze. About the same time our force engaged the enemy and drew part of the attention they were giving the Rough Riders. This, the latter claimed, enabled them to continue the movement on the enemy's works.

But as our command had an equal number of 1st and 10th Cavalrymen, I am of the opinion that the story of our saving the Rough Riders arose from the fact that as soon as the fight was over, the 1st Regular Cavalry was opening its arms to us, declaring that we, especially B Troop, had saved them; for the 1st Regular Cavalry was first in the attack in General Young's command; and when the enemy began to make it pretty warm, he ordered B and I Troops of the 10th forward on the right.

Troop B was in the lead; and the alacrity with which these two troops moved to the front has always been praised by the 1st Cavalry; and they declare that that movement helped them wonderfully. In making this movement my troop had three or four men wounded; and later, when Sergeant Thompson's squad was fighting far to the front, it had in it several members of the 1st Cavalry, who are always glad to praise him.

So, I think that by the Rough Riders first attributing their success, or their rescue from inevitable defeat, to the attack made by our command; and by the 1st Regular Cavalry's very generously, in the heat of success, bestowing upon us the honors of the day, it finally became a settled thing that we saved the whole battle.

That evening, after the battle, I was met by Lieutenant Shipp, later killed at San Juan Hill, who, on inquiring and being told that I belonged to Troop B, congratulated me on its conduct, and said it had made a name for the regiment. Lieutenant Shipp was not in that fight, but had come up after it was over and had heard of us through the 1st Cavalry."

C

Sergeant John Buck was born September 10th, 1861, at Chapel Hill, Texas; enlisted in 10th Cavalry, November 6, 1880, and passed over ten years in active Indian service. He is a man of strong character, an experienced horseman and packer, and so commanded a portion of the firing line in the battle of June 24 as to elicit remarks of praise from officers of other troops "for his gallantry, coolness and good judgment under fire." Sergeant Thompson's good conduct in the same battle was noticeable also. Sergeant Buck was made second lieutenant in the 7th U. S. Volunteer Infantry and subsequently captain in the 48th United States Volunteers.

CHAPTER VI.

THE BATTLE OF EL CANEY.

The Capture of the Stone Fort by the Twenty-fifth Infantry.

While the battle of Guasimas was going on, in which the Tenth Cavlary took so conspicuous a part, the Twenty-fourth Infantry still remained on board the City of Washington awaiting orders to land. During the night of the 24th such orders were received by the authorities of the transport, and they were directed to land their troops, but the General Commanding, Brigadier-General Kent, did not hear of the matter until some time the next morning. He relates the following circumstances in his official report of the debarkation:

"At 9 a. m. of the 25th Lieutenant Cardin, of the Revenue Marine, came aboard with orders for me to proceed to and disembark at Altares (Siboney). This officer also handed me a letter from the corps commander expressing his astonishment that I had remained away three days."

General Kent also states in his report that his travel rations had been exhausted seven days before and that but one meal of field rations remained, and that the ship's supply both of water and provisions was running low, and that in consequence of these facts as well as for higher considerations he was very anxious to get on shore. The debarkation followed as rapidly as possible, and that afternoon General Kent reported in person to Major-General Wheeler, the troops bivouacking for the night near the landing. The next day Colonel Pearson, who commanded the Second Brigade of Kent's division, took the

Second Infantry and reconnoitred along the railroad toward the Morro, going a distance of about six miles and returning in the evening, having found no enemy in that vicinity, although evidences were found that a force had recently retreated from a blockhouse situated on the railroad about two miles from Aguadores.

On the day following, June 27th, the entire division moved out on the road toward Santiago and encamped on the same ground that Lawton had occupied the night previous. The Second Brigade took its place near Savilla, while the Third Brigade, which included the Twenty-fourth Infantry, went into camp at Las Guasimas, where the affair of the 24th had occurred. The order of march had now partially fallen back to the original plan: Lawton in advance, with whom was the Twenty-Fifth Infantry; Wheeler next, with whom was the Ninth and Tenth Cavalry, and Kent in the rear, who had, as we have just related, the Twenty-fourth Infantry in his Third Brigade. In this order the army moved, so far as it moved at all, until the morning of the 30th, when dispositions for the general attack began.

The story of the great battle, or as it turned out, of the two great battles, begins on this day, and the careers of the four colored regiments are to be followed through the divisions of Lawton, Kent and Wheeler. Let us begin, however, with General Shafter's official report and his "Story of Santiago," as told in the "Century" of February, 1899.

From these sources it is learned that on June 30th General Shafter reconnoitered the country about Santiago and determined upon a plan of attack. Ascending a hill from which he could obtain a good view of the city, and could also see San Juan Hill and the country about El Caney, he observed afresh what had impressed itself upon all immediately upon landing, to wit: That in all this country there were no good roads along which to move troops or transport supplies. The General says: "I had never seen a good road in a Spanish country, and Santiago did not disappoint my expectations." The roads as he saw them from the summit of the hill on June 30th were very poor, and indeed, little better than bridle paths, except between El Caney and San Juan River and the city. Within this region, a distance of from four to four and a half miles, the roads were passable. El Caney lay about four miles northeast of Santiago, and was strongly fortified, and, as events proved, strongly garrisoned. This position was of great importance to the enemy, because from it a force might come to attack the right flank and rear of the American Army as it should make its attempt on San Juan Hill. El Caney held the road from Guantanamo, at which point an important Spanish force was posted. While General Shafter was surveying the country from the hill at El Pozo and making what special examination he could of the country toward San Juan Hills, Generals Lawton and Chaffee were making a reconnoisance around El Caney. From General Lawton's report it would appear that the work of reconnoitering around El Caney was done chiefly by General Chaffee. He says: "To General Adna R. Chaffee I am indebted for a thorough and intelligent reconnoissance of the town of El Caney and vicinity prior to the battle and the submission of a plan of attack which was adopted. I consider General Chaffee one of the best practical soldiers in the army and recommend him for special distinction for successfully charging the stone fort mentioned in this report, the capture of which practically closed the battle."

The general plan of attack as explained by General Shafter

himself in his "Century" article was "to put a brigade on the road between Santiago and El Caney, to keep the Spaniards at the latter place from retreating on the city, and then with the rest of Lawton's division and the divisions of Wheeler and Kent, and Bates' brigade to attack the Spanish position in front of Santiago." Before that he had said that he wished to put a division in on the right of El Caney and assault the town on that road. To Admiral Sampson on June 26th he said: "I shall, if I can, put a large force in Caney, and one perhaps still farther west, near the pipe-line conveying water to the city, making my main attack from the northeast and east." His desire at this time was to "get the enemy in my front and the city at my back." On June 30th he had modified this plan so as to decide to place one brigade on the road between El Caney and Santiago, with a view merely to keeping the El Caney garrison from retreating into Santiago.

As he was explaining his plan to the division officers and others on the afternoon of the 30th at his own headquarters, Lawton and Chaffee were of the opinion that they could dispose of the Spaniards at El Caney in two hours time. "Therefore," says the General, "I modified my plan, assigning Lawton's whole division for the attack of El Caney and directed Bates' independent brigade to his support." This last modification of General Shafter's plan was made in deference to the opinion of subordinates, and was based upon observations made especially by General Chaffee.

The force assigned for the reduction of El Caney was to begin its work early in the morning, and by ten or eleven o'clock at the outside it was expected that the task would be accomplished and Lawton would join Kent and Sumner in the assault upon San Juan. Early on the morning of July 1st Capron's battery was got into position on a line running directly north from Marianage on a hill about five hundred yards east of Las Guasimas Creek. Lawton's division began its move on the afternoon of the 30th, as did in fact the whole army, and bivouacked that night near El Pozo. The Twenty-fifth Infantry, which belonged to the Second Brigade, commanded by Colonel Miles, a former Major of the Twenty-fifth, left El Pozo at daylight by way of the road leading almost due north, and marched about one mile to the little town of Marianage. Here a halt was made for an hour, from 6.30 to 7.30, during which time reconnoitering parties were sent out to examine the ground toward the Ducoureau House, which lay about one mile to the northward of Marianage, and which had been designated by General Lawton as a general rendezvous after the engagement should terminate. Reconnoissance was made also to the front for the purpose of discovering the enemy, and to ascertain the left of Ludlow's brigade. This was the first brigade of Lawton's division and consisted of the Eighth and Twenty-second Infantry and the Second Massachusetts, the last named regiment being on the right. The Second Brigade was to connect with this on its right and succeeded in finding the position of the Second Massachusetts during this halt. At 11.30 Miles' brigade was ordered to take position on the right of Ludlow's brigade, which it did in the following order: The Fourth Infantry on the left, joining with the Second Massachusetts on Ludlow's right; the Twenty-fifth on the right, with its left joining on the Fourth Infantry.

We must now review the progress of the battle so far as it is possible to do so, from the firing of the first shot by Capron's battery up to 11.30, an hour long after the time at which it had been supposed that El Caney would fall. Capron's reports are very brief. He says: "July 1—Fired shell and shrappnel into El Caney (ranged 2.400) 6.15 a. m. to 11.30 a. m." In another report he says: "Opened fire July 1, with shell and shrappnel at 6.15 on Caney; range, 2,400 yards; continued until 11.30 a. m." He says that the battery "continued its fire against specified objectives intermittently throughout the day under the personal direction of the division commander." The forces we have so far considered, consisting of Ludlow's and Miles' brigades, and of Capron's battery, lay to the south of Caney, between it and Santiago, Ludlow's brigade having been placed there to "cut off the retreat of the garrison should it attempt to escape." Up to 11.30 there had been no call for employing it for that purpose. The garrison had made no attempt to escape. We must now go around to the east and north of Caney. Here the Third Brigade, consisting of the Seventh, Twelfth and Seventeenth Infantry, was posted, and early in the morning joined in the attack, the brigade getting under fire before eight o'clock. Colonel Carpenter, of the Seventh Regiment, says that one company of his regiment, by General Chaffee's direction, was detached and sent forward to reduce a blockhouse, well up on the hill, which commanded the approach of his regiment to the field of action. After several ineffectual attempts by the company, the Captain (Van Orsdale) was directed to abandon the undertaking and rejoin the regiment, which then took up a position on the crest of a hill running nearly parallel with the Spanish lines. From this position the men crawled forward about fifty yards and opened a deliberate fire upon the enemy, keeping it up for about an hour, but as the losses of the regiment at this time were considerable and the fire seemed to be without material effect, the command was withdrawn to its position on the hill where it found protection in a sunken road. In this condition this regiment lay when Capron's battery made its lull at 11.30. The fearful fire this regiment met can be estimated by the losses it sustained, which during the day were as follows: Killed ,1 officer and 33 enlisted men; wounded, 4 officers and 95 enlisted men; missing, 3 enlisted men. The Seventeenth Regiment went into action on the right of the Seventh, doing but little firing, as their orders were not to open fire unless they could make the fire effective. Companies C and G fired a few volleys; the remainder of the regiment did not fire at all. Four enlisted men were killed and two officers severely wounded, one, Lieutenant Dickinson, dying from his wounds within a few hours. Several enlisted men were also wounded. At 11.30 this regiment was lying on the right of the Seventh. The Twelfth Regiment began firing between 6 and 7 in the morning and advanced to take its position on the left of the Seventh Infantry. This regiment early reached a position within 350 yards of the enemy, in which it found shelter in the sunken road, "free from the enemy's fire." The regiment remained in this position until about 4 o'clock in the afternoon, and, hence, was there at 11.30 a. m. The losses of this regiment during the day were, killed, 7 enlisted men; wounded, 2 officers and 31 enlisted men. From these brief sketches the reader will now be able to grasp the position of Lawton's entire division. Beginning on the south, from the west, with Ludlow's brigade, consisting of the Twenty-second, Eighth and Second Massachusetts, the line was continued by Miles' brigade of the Fourth and Twenty-fifth Infantry; then passing over a considerable space, we strike Chaffee's brigade, posted as has just been described. General Bates' brigade probably arrived upon the field about noon. This brigade consisted of the Third and Twentieth Infantry, and is known as "Bates' Independent Brigade." The brigade is reported as going into action about 1 o'clock and continuing in action until 4 o'clock. It took a position on the right, partially filling up the gap between Miles and Chaffee. The first battalion of the Twentieth Infantry went into action on the left of the Twenty-fifth Infantry's firing line, and one company, A, took part in the latter part of the charge by which the stone house was taken. Between 11.30, when Capron's firing stopped, and when Miles' brigade was moved forward to join the right of Ludlow's, and 12.20, when the battery recommenced, the troops, including Bates' brigade, were either in the positions described above or were moving to them. Noon had arrived and El Caney is not taken; the garrison has not attempted to escape, but is sending out upon its assailants a continuous and deadly fire. "Throughout the heaviest din of our fire," says Colonel Carpenter, "could be heard the peculiar high-keyed ring of the defiant enemy's shots."

Twelve o'clock on July 1st, 1898, was a most anxious hour for our army in Cuba. The battle at El Caney was at a standstill and the divisions of Kent and Sumner were in a most perilous situation. Bonsal's description of the state of the battle at that time is pathetic. Speaking of the artillery at El Caney —Capron's battery—he says it was now apparent that this artillery, firing from its position of twenty-four hundred yards, could do very little damage to the great stone fort and earthworks north of the village. The shots were too few and the metal used too light to be effectual. Three hours of the morning had worn away and the advance of our men had been slowly made and at great cost; all the approaches were commanded by Spanish entrenchments and the fighting was very unequal. A soldier of the Twenty-fifth says that when he came in sight of the battle at El Caney, "the Americans were gaining no ground, and the flashes of the Spanish mausers told us that the forces engaged were unequally matched, the difference of position favoring the Spaniards." This view was had about noon, or soon after. At that time "a succession of aides and staff officers came galloping from headquarters with messages which plainly showed that confusion, if not disaster, had befallen the two divisions which, by the heavy firing, we had learned to our great surprise, had become warmly engaged in the centre. The orders to General Lawton from headquarters were at first peremptory in character—he was to pull out of his fight and to move his division to the support of the centre" (Bonsal). This call for Lawton arose from the fact that about noon General Shafter received several dispatches from Sumner, of the Cavalry Division, requiring assistance. General Sumner felt the need of the assistance of every available man in the centre of the line where he was carrying on his fight with the Spaniards on Blue House Hill. This situation so impressed the General, Shafter, that he finally wrote to Lawton, "You must proceed with the remainder of your force and join on immediately upon Sumner's right. If you do not the battle is lost." Shafter's idea then was to fall back to his original plan of just leaving enough troops at El Caney to prevent the garrison from going to the assistance of any other part of the line. Shafter himself says: "As the fight progressed I was impressed with the fact that we were meeting with a very stubborn resistance at El Caney and I began to fear that I had made a mistake in making two fights in one day, and sent Major Noble with orders to Lawton to hasten with his troops along the Caney road, placing himself on the right of Wheeler"

(Sumner). Lawton now made a general advance, and it is important to see just what troops did advance. The Seventh Infantry did not move, for Lieutenant-Colonel Carpenter says that after withdrawing "to the partial cover furnished by the road, the regiment occupied this position from 8 o'clock a. m. until about 4.30 p. m." The Seventeenth did not move, for Captain O'Brien, commanding, says the regiment took a position joining "its left with the right of the Seventh Infantry" and that the regiment "remained in this position until the battle was over." The Twelfth Infantry remained in its shelter within 350 yards of the stone fort until about 4 p. m. Then we have Chaffee's brigade on the north of the fort remaining stationary and by their own reports doing but little firing. The Seventeenth fired "for about fifty minutes" about noon, with remarkable precision, but "it seemingly had no effect upon reducing the Spanish fire delivered in our (their) front." The Seventh did not fire to any extent. The Twelfth Infantry lay in its refuge "free from the enemy's fire" and may have kept up an irregular fire.

About this time Bates' brigade entered the field and one battalion of the Twentieth Infantry is reported to have joined the left of the firing line of the Twenty-fifth. General Ludlow says there was a lull from 12 to 1 p. m., "when the action again became violent, and at 3 p. m. the Third Brigade captured the stone fort with a rush and hoisted the American flag." From Ludlow's brigade, Captain Van Horne, commanding the Twenty-second Infantry, after the wounding of Lieutenant-Colonel Patterson, says that the First Battalion of his regiment took a position about 800 yards from the town and kept up firing until the place surrendered. He does not say positively that the firing was upon the town, but he had said just before that the Second Battalion slowly moved forward, firing into the town from the left, so that we may readily conclude from the context as well as from the position that the First Battalion fired into the town also. Hence it seems fair to exclude from the fort all of Ludlow's brigade, and it is observable that Ludlow himself claims no part in the capture of that stronghold.

General Bates says his brigade took position to the right of Colonel Miles' brigade and pushed rapidly to the front. He then says that after remaining sometime in the crossroad to the right of Miles' brigade, under a heavy fire from the enemy, the brigade moved farther "to the right to the assault of a small hill, occupied upon the top by a stone fort and well protected by rifle pits. General Chaffee's brigade charged them from the right, and the two brigades joining upon the crest, opened fire from this point of vantage, lately occupied by the Spanish, upon the vilalge of El Caney." General Chaffee says it was in consequence of the fire of General Bates' troops upon the fort that the assault by the Twelfth Infantry was postponed.

In General Chaffee's report this statement occurs: "The action lasted nearly throughout the day, terminating at about 4.30 p. m., at which time the stone blockhouse was assaulted by Captain Haskell's battalion of the Twelfth Infantry, under the personal direction of Lieutenant-Colonel Comba, commanding the regiment. The resistance at this point had been greatly affected by the fire of Capron's battery. A few moments after the seizure of this point—the key to the situation—my left was joined by General Bates with a portion of his command." It is to be noted in connection with all of the above statements that Major McCaskey, who commanded the Twentieth Infantry (Bates brigade), says: "The First Battalion was moved to the right and put into action on the left of the

Twenty-fifth Infantry's firing line, and one company, A, took part in the latter part of the charge by which the stone house was taken." The two points to be noted here are (1) that this battalion was on the left of the Twenty-fifth's firing line, and (2) that one company took part in the charge upon the stone house. When Chaffee's brigade charged the stone house from the right some of Bates' troops, at least this Company A, from the battalion near the firing line of the Twenty-fifth Infantry, took part in the latter part of the charge. The two brigades, Bates' and Chaffee's, joined immediately after the capture of the stone fort and opened fire upon the town.

We have now traced the actions and the fortunes of the three following brigades: Ludlow's Chaffee's and Bates'. But what has become of Miles' brigade? Unfortunately, the Second Brigade has not been so well reported as were the others engaged in the action at El Caney. We have seen that it was ordered to take position on the right of Ludlow's brigade at 11.30, when Capron's battery ceased its firing for the fifty minutes. "We were detained in reaching our position by troops in our front blocking the road," says the brigade commander. "We came into action directly in front of the stone blockhouse at 12.30, and from that hour until about 4.30, when the command 'cease firing' was given, the blockhouse having been captured, my command was continuously under fire." The reader will note in this report that the brigade went into action at 12.30, several hours before the charge was ordered by General Chaffee, and at least an hour and a half before, according to the report of the commander of the Third Brigade, "this fort was practically in the possession of the Twelfth Infantry." Major Baker, who commanded the Fourth Infantry, says: "About 12 m. we received orders directing us to take our place in the line of battle, and arriving at the proper point the regiment was placed in line in the following order: The First Battalion in the fighting line; the Second Battalion in support and regimental reserve. In this order the First Battalion, under my command, took up the advance toward the blockhouse, to our right, south east of Caney." This battalion advanced until it reached a position about 200 yards from the village, where it remained, assisted by the Second Battalion until the capture of the fort. Two companies of this First Battalion "fired into the town and also into the blockhouse until its fall." A good part of the fire of this regiment was directed upon the fort.

Colonel Miles says: "The brigade advanced steadily, with such scanty cover as the ground afforded, maintaining a heavy fire on the stone fort from the time the fight began until it ended." The reader is asked to note particularly that this fire was continuous throughout the fight; that it was characterized by the brigade commander as "heavy," and that it was "on the stone fort.". He says: "As the brigade advanced across a plowed field in front of the enemy's position the latter's sharpshooters in the houses in Caney enfiladed the left of our line with a murderous fire. To silence it Major Baker, Fourth Infantry, in command of the battalion of that regiment on the left of our line of battle, directed it to turn its fire upon the town. In so doing this battalion lost heavily, but its steady front and accurate volleys greatly assisted the advance of the remainder of the brigade upon the stone fort."

We have now these facts clearly brought out or suggested: That the brigade took its place in line of battle soon after 12 o'clock; that the Fourth Infantry was on the left; that the advance of the First Battalion of the Fourth Infantry was "toward the blockhouse;" that aside from the companies of the

Fourth Infantry that fired into town, "the remainder of the brigade advanced upon the stone fort." The Fourth Infantry, holding the left of the line, however, reached a position from which it could not advance, its commander having "quickly perceived that an advance meant annihilation, as it would involve not only a frontal, but also a flank fire from the town." Here the Fourth Infantry remained, but continued to maintain a fire upon both the blockhouse and the town.

There is but one more regiment in all of Lawton's division to be accounted for, and that is the Twenty-fifth Infantry, holding the right of Miles' brigade in this advance. This regiment was in place in the line under its gallant and experienced commander, Lieutenant-Colonel A. S. Daggett, and contributed its full share of that "heavy fire on the stone fort from the time the fight began until it ended." Major McCaskey says the First Battalion of his regiment took a position on the left of the Twenty-fifth's firing line. The statement seems erroneous, and one is inclined to believe that it was originally written "on the right," instead of "on the left"; but it is enough for our purpose now, that the firing line of the Twenty-fifth is recognized well in advance. Major Baker, who commanded on the left of the brigade line, and whose advance was stopped by the flank fire from the village and a frontal fire from the fort, says: "as a matter of fact the village of El Caney was not charged by any troops. Those of Bates' brigade and the Twenty-fifth Infantry, after having carried the stone fort (on a hill some 75 feet higher, and to the east of the town,) fired into the village, and the Fourth Infantry continued its fire. Nor was it charged by any of the troops to our left. Such a charge would necessarily have been seen by us." Major Baker, who was on the field and had the blockhouse in clear view, declares that some of Bates' brigade and the Twenty-fifth Infantry carried the stone fort. Major McCaskey says that one battalion of the Twentieth Infantry (Bates' brigade) was on the left of the Twenty-fifth's firing line, and that one company (A) took part in the latter part of the charge by which the fort was taken. This battalion may be referred to by Major Baker when he says: "Those of Bates' brigade and the Twenty-fifth Infantry, after having carried, etc."

As there are some matters of dispute concerning the events which I am now going to relate, I will present a soldier's statement before I go to the official records. The soldier in writing to me after the battle says: "I was left-guide of Company G (25th Infantry), and I received orders from Lieutenant McCorkle to guide on Fourth Infantry, which held the left flank. 'Forward, march! Guide left. Don't fire until you see somebody; then fire to hit!' came the orders. Tramp! tramp! Crash! crash! On we walked and stopped. We fired into the underbrush for safety; then in another moment we were under Spanish fire. Balls flew like bees, humming as they went. Soon we found ourselves up against a network of Spanish trickery. Barbed-wire fences, ditches and creeks, too numerous to think of. The only thing left was to go ahead or die; or else retreat like cowards. We preferred to go ahead. At this first fence Lieutenant McCorkle was taken to earth by a Spanish bullet. Lieutenant Moss spoke out, 'Come ahead! Let's get at these Spaniards!' A few moments more and he, too, was almost dead with exertion, loud speaking, running and jumping, as onward we swept toward the Spanish stronghold. The sun was exceedingly hot, as on the slope of a little mound we rested for a few moments. We lay here about five minutes, looking into the Spanish fort or blockhouse; we measured the distance by our eyesight, then with our rifles; we began to cheer and storm, and in a moment more, up the hill like a bevy of blue birds did the Twenty-fifth fly. G and H Companies were the first to reach the summit and to make the Spaniards fly into the city of El Caney, which lay just behind the hill. When we reached the summit others soon began to *mount our ladder.* We fired down into the city until nearly dusk."

The brigade made its advance under fire almost from the beginning. The commander says it was continuously under fire from 12.30 to 4.30 p. m. "The attack was begun by two companies in each regiment on the firing line, strengthened by supports and reserves from the remaining companies until the brigade had but two companies in reserve. At one time in this hotly engaged contest the commanding officer of the Twenty-Fifth Infantry sent me word that he needed troops on his right. I then sent forward 40 Cubans, under command of Captains Jose' Varges and Avelens Bravo, with Lieutenants Nicholas Franco and Tomas Repelao, to form on the right of the Twenty-fifth, which was also the right of the brigade. With these Cubans I ordered Private Henry Downey, Company H, First Infantry, on duty as interpreter at the headquarters. These men advanced on the stone fort with our line, fighting gallantly, during which Lieutenant Nicholas Franco was mortally wounded and died soon afterwards." (Col. Miles' report.)

From the soldier's story, as well as from the official report of the brigade commander, it is conclusive that the real objective of the Second Brigade was the stone fort, and that the Twenty-fifth Infantry, which occupied the right of the line, had no other objective whatever.* It also appears that Bates' brigade, although somewhere on the right, was not so near but that the commanding officer of the Twenty-fifth could see the need of troops at his right; and to meet this need the brigade commander "sent forward 40 Cubans, who advanced on the stone fort with our lines." The fire from this fort continued severe during the whole of the advance, and until the last halt made by the Twenty-fifth. At the first fence met by the Twenty-fifth Lieutenant McCorkle was killed; and, to use the words of a soldier, "as the regiment swept toward the Spanish stronghold" to reach the slope of a little mound for cover, many more fell. Behind this little mound, after resting about five minutes, they began their last fire upon the enemy. This must have been as late as 3 o'clock, and perhaps considerably later, and the fire from the stone fort was vigorous up until their last halt, as their casualties prove. The battery had begun to fire on the fort again at 12.30 and continued from the same position until 2.10, the range being as has been already stated, 2,400 yards. Hence the artillery firing at long range had ceased, and it is generally conceded that this long range firing had been ineffective. Captain Capron says he moved his battery at 2.10 p. m. to 1,000 yards from Caney and opened fire on two blockhouses. He does not say at what hour he opened fire on these two blockhouses, or how long he continued to fire, or what was the effect of his fire upon the two block houses. Lieutenant-Colonel Bisbee, who was acting as support of Capron's battery, says of himself that he "moved with the battery at 3.30 p. m. by the Dubroix (Ducureaux) road." General Lawton says the battery was moved to a new position about 2.30, "about 1,000 yards from certain blockhouses in the town, where a few shots, all taking effect, were fired." From these reports it would appear that after moving to the second station the battery fired upon two blockhouses in the town, and not upon the stone fort. General Ludlow, speaking of the battle, says: "In the present case, the artillery fire was too distant to reduce the blockhouses or destroy the entrenchments, so that the attack was practically by infantry alone." On the other hand, General Chaffee says: "The resistance at this point," meaning the stone fort at the time of assault, "had been greatly affected by the fire of Capron's battery." Colonel Comba, of the Twelfth Infantry, says: "The artillery made the breach through which our men entered the stone work." Bonsal says that Captain Capron, "under the concentrated fire of his four guns at a point blank range of a thousand yards, had converted the fort into a shapeless ruin," when the infantry charged it.

It is probable that in this case, as in most cases of similar nature, the truth divides equally between the apparently opposing views. Of General Ludlow, who is the authority for this statement, that the stone fort at El Caney was taken by infantry alone, General Lawton says: "General Ludlow's professional accomplishments are well known and his assignment to command a brigade in my division I consider a high compliment to myself." "The fighting was all done with small arms" were the words written me by an infantryman soon after the battle. The question, whether Capron fired upon the stone fort after taking his new position, or fired on two blockhouses, entirely distinct from the fort, remains undetermined. The author of this work inclines to the conclusion that the fire of Capron after moving to his new position was directed for a brief period, at least, upon the stone fort.

Inasmuch as we are now to trace the career of the Twenty-fifth Infantry through an unfortunate dispute, on both sides of which are officers of high rank and unimpeachable honor, it is important to note, first, to what extent the several statements, both unofficial and official, can be harmonized and made to corroborate one another. Major Baker says: "Those of Bates' brigade and the Twenty-fifth Infantry, after having carried the stone fort," which he explains was some 75 feet higher than the town, then fired *down* into the village. The soldier who acted as left-guide of Company G, Twenty-fifth Infantry, says, after getting up on the hill, "we fired *down* into the city until near dusk." The experience of the soldier agrees exactly with the report of the officer. The fact that the Twenty-fifth went up the hill cannot be questioned, and that up to their last halt, they went under fire, no one will deny. Bonsal, in speaking of Chaffee's brigade, which was "more immediately charged with the reduction of Caney" (Ludlow's report), says: "And it was nearly five o'clock when his most advanced regiment, the gallant Twelfth Infantry, deployed into the valley and charged up the steep hillside, which was lined with Spanish trenches, rising in irregular tiers and crowned with a great stone fort." The stone fort at this time, however, was, as he says, "a shapeless ruin." Where was the Twenty-fifth Infantry at this time? Mr. Bonsal continues: "Almost at the same moment the Twenty-fifth Colored Infantry, the leading regiment of Miles' brigade, which had been advancing in the centre, started up the hill also." General Lawton says that after moving the battery to its new position, 1,000 yards from certain blockhouses in the town, Capron fired a few shots, all of which took effect, and he adds: "This firing terminated the action, as the Spanish garrison were attempting to escape." Colonel Comba says there was a breach in the stonework large enough for his men to enter, and that this had been made by the artillery; General Chaffee says resistance had been greatly affected by the artillery, and Bonsal adds, the garrison resisted the last advance made by the infantry but for a moment.

General Chaffee declares: "The troops arriving at the fort were there in the following order: Twelfth Infantry, which took the place; the command of General Bates some moments later; the Twenty-fifth Infantry."

The facts therefore stand, that the Twenty-fifth Infantry was on the ground with the first troops that reached the fort and that there was a captain of that regiment who then and there claimed the capture of the place, even against the claims of a Major-General. He was told that his proposition was absurd, and so it may have been from one standpoint; and yet there may be a ground upon which the captain's claim was fair and just.

That the Twelfth Infantry arrived on the ground first is not disputed; but it is questioned whether the fort was belligerent at that time. General Chaffee says the resistance had been greatly reduced by the artillery; General Lawton says the action had been finished by Capron's shots and the garrison was trying to escape; a soldier from the Twenty-fifth says the Spaniards flew out of the fort to the town; Bonsal says, they stoutly resisted "for a moment and then fled precipitately down the ravine and up the other side, and into the town." If first occupancy is the only ground upon which the capture of a place can be claimed, then the title to the honor of capturing the stone fort lies, according to official report as so far presented, with the Twelfth Infantry. But even upon this ground it will be shown that the Twenty-fifth's action will relieve the claim of its captain from absurdity. We are now prepared to read the official report of the commanding officer of the Twenty-fifth Regiment, Lieutenant-Colonel Daggett, who was with the regiment all through the fight, and who bore himself so well that the division commander said: "Lieutenant-Colonel Daggett deserves special mention for skillful handling of his regiment, and would have received it before had the fact been reported by his brigade commander."

July 5, 1898.

Intrenchments Twenty-fifth United States Infantry,
Adjutant-General, Second Brigade, Second Division, Fifth Corps.

Sir:—I have the honor to submit the following report of the part taken by the Twenty-fifth Infantry in the battle of the 1st instant. The regiment formed firing line on the right of the Fourth Infantry, facing a Spanish fort or blockhouse about half a mile distant. On moving forward, the battalion, composed of Companies C, D, E, G and H, and commanded by Capt. W. S. Scott, received the fire of the enemy, and after advancing about 400 yards was subjected to a galling fire on their left. Finding cover, the battalion prepared for an advance up the hill to the fort. This advance was made rapidly and conducted with great skill by company officers.

"On arriving within a short distance of the fort the white flag was waved to our companies, but a cross fire prevented the enemy from advancing with it or our officers from receiving it. About twenty minutes later a battalion of some other regiment advanced to the rear of the fort, completely covered from fire, and received the flag; but the men of the Twenty-fifth Infantry entered the fort at the same time. All officers and men behaved gallantly. One officer was killed and three wounded; eight men were killed and twenty wounded.

"About 200 men and ten officers were in the firing line. I attribute the comparatively small losses to the skill and bravery of the company officers ,viz.: First Lieutenant Caldwell and Second Lieutenants Moss and Hunt. Second Lieutenant French, adjutant of the battalion, was among those who gallantly entered the fort.

"The battle lasted about two hours and was a hotly contested combat.

Very respectfully,

"A. S. DAGGETT,

"Lieutenant-Colonel, Twenty-fifth Infantry, Commanding."

Here it is shown by the testimony of the regimental commander, that a battalion of the Twenty-fifth ascended the hill and arrived at a short distance from the fort about twenty minutes before any other troops are mentioned as coming in sight; and that a white flag was waved to the companies of the Twenty-fifth. It was doubtless upon this ground that a captain of the Twenty-fifth had the temerity to claim the capture of the place, even from a Major-General. I do not know who the captain was, but it is evident that he had what he believed ample grounds for his claim. Colonel Daggett says, also, that when the men of the other regiment advanced to this fort after it had waved the white flag to the companies of the Twenty-fifth, the men of the Twenty-fifth advanced and entered the fort at the same time. Bonsal says: "Almost at the same moment that the Twelfth started up the hill the Twenty-fifth started up the hill also;" while according to Colonel Daggett's testimony the Twenty-fifth was well up the hill already and the fort had waved to it the white flag.

Colonel Daggett makes this further report:

Headquarters Twenty-fifth Infantry,
Near Santiago, Cuba, July 16, 1898.

The Adjutant-General, Second Division, Fifth Corps, near Santiago, Cuba.

Sir:—Feeling that the Twenty-fifth Infantry has not received credit for the part it took in the battle of El Caney on the first instant, I have the honor to submit the following facts:

I was ordered by the brigade commander to put two companies (H, Lieutenant Caldwell, and G, Lieutenant McCorkle) on the firing line in extended order. The right being uncovered and exposed to the enemy, I ordered D Company (Captain Edwards) to deploy as flankers. The battalion was commanded by Capt. W. S. Scott. The battalion advanced about 300 yards under fire, the Fourth Infantry on its left, where the line found cover, halted, and opened fire on the blockhouse and intrenchments in front of it. After the line had been steadied and had delivered an effective fire, I ordered a further advance, which was promptly made. As the Fourth Infantry did not advance, my left was exposed to a very severe fire from the village on the left. I immediately ordered Company C (Lieutenant Murdock), which was in support, to the front, and E. Company (Lieutenant Kinnison) from regimental reserve to take its place. Thus strengthened, the four companies moved up the hill rapidly, being skilfully handled by company officers. On arriving near the fort the white flag was waved toward our men, but the fire from the village on our left was so severe that neither our officers nor Spanish could pass over the intervening ground. After about twenty minutes some of the Twelfth Infantry arrived in rear of the fort, completely sheltered from the fire from the village, and received the white flag; but Privates J. H. Jones, of Company D, and T. C. Butler, H. Company, Twenty-fifth Infantry, entered the fort at the same time and took possession of the Spanish flag. They were ordered to give it up by an officer of the Twelfth United States Infantry, but before doing so they each tore a piece from it, which they now have. So much for the facts.

I attribute the success attained by our line largely to the bravery and skill of the company officers who conducted the line to the fort. These officers are: First Lieutenants V. A. Caldwell and J. A. Moss, and Second Lieutenant J. E. Hunt. It is my opinion that the two companies first deployed could not have reached the fort alone, and that it was the two companies I ordered to their support that gave them the power to reach it. I further believe that had we failed to move beyond the Fourth Infantry the fort would not have been taken that night.

The Twenty-fifth Infantry lost one officer killed* and three wounded, and seven men killed and twenty-eight wounded.

Second Lieutenant H. W. French, adjutant of Captain Scott's battalion, arrived at the fort near the same time as the other officers.

I request that this report be forwarded to corps headquarters.

Very respectfully,

A. S. DAGGETT,

Lieutenant-Colonel, Twenty-fifth Infantry, Commanding.

*First Lieutenant McCorkle killed; Captain Edwards and First Lieutenants Kinnison and Murdock wounded.

General Chaffee's statement is not to be questioned for a moment. There is not the least doubt that the troops, as organizations arrived at the fort in the order he describes. General Lawton says: "General Chaffee's brigade was especially charged with the duty of assaulting the stone fort, and successfully executed that duty, after which a portion of the Twenty-fifth, and a portion of Bates' brigade, assisted in the work, all of which is commendable." He says also, that the "Twenty-fifth Infantry did excellent service, as reported, though not better than the others engaged.' This seems to confirm Lieutenant-Colonel Daggett's report, for he says he is sure the regiment did excellent work, "as reported;" and at that time he is commenting on Lieutenant-Colonel Daggett's report, the report printed above. The broad statements of General Lawton do not touch the exact question at issue between the reports of the subordinate commanders; nor do they throw any light on the circumstances of the final charge. Miles' brigade had been advancing on the stone fort for some hours, and the Twenty-fifth was so near when the charge of the Twelfth was made that portions of it were on the hill and near the fort at the same time. The commander of the Third Brigade saw the fight from one side and reported events as he learned them. His official statement requires no support. The commanding officer of the Twenty-fifth Infantry saw the fight from another standpoint, and his official reports are entitled to equal respect. Both the General's and the Lieutenant-Colonel's must be accepted as recitals of facts, made with all the accuracy that high personal integrity armed with thorough military training can command. Happily the statements, which at first appear so widely at variance, are entirely reconcilable. The following supplementary report of the regimental commander, when taken in connection with the final complimentary orders published in the regiment before leaving Cuba, will place the whole subject before the reader and put the question at rest, and at the same time leave undisturbed all the reports of superior officers.

Headquarters Twenty-fifth Infantry,
Montauk Point, Long Island, August 22, 1898.

The Adjutant-General, U. S. Army, Washington, D.C.

Sir:—I have the honor to submit a supplementary report to the original one made on the 19th (16th) of July, 1898, of the battle of El Caney de Cuba, so far as relates to the part taken therein by the Twenty-fifth Infantry:

1. I stated in the original report that the Twenty-fifth Infantry, in advancing, broke away from and left the Fourth Infantry behind. This may inferentially reflect on the latter regiment. It was not so intended, and a subsequent visit to the battle-field convinces me that it would have been impossible for the regiment to advance to the fort, and, although it might have advanced a short distance farther, it would have resulted in a useless slaughter, and that the battalion commander exercised excellent judgment in remaining where he did and by his fire aiding the Twenty-fifth Infantry in its advance.

2. Colonel Miles, the then brigade commander, informed me that his first report of the battle would be brief and that a later and full report would be made. In his former report I think he failed to give credit to myself and regiment. As he was soon after relieved of the command of the brigade I assume that no further report will be made.

I have reported what the regiment did, but said nothing about my own action. I must, therefore, report it myself or let it go unrecorded. Distasteful as it is to me, I deem it duty to my children to state the facts and my claims based thereon, as follows:

1. I was ordered to put two companies in the firing line. Before this line advanced the brigade commander informed me, and personal examination verified, that my right was in the air and exposed. On my own judgment I ordered a company, as flankers, to that part of the line.

2. As soon as the line had rested and become steadied at its first halt I ordered it to advance, and it continued to advance, although it broke away from the rest of the brigade.

3. As this exposed the left to a galling and dangerous fire, I ordered, on my own judgment, a company to re-enforce that part of the line and a company from the regimental reserve also to the fighting line.

These are the facts, and as my orders were to keep my left joined to the right of the Fourth Infantry, and received no further orders, my claims are as follows:

1. That it was necessary to place a company on the right as flankers.

2. That the conditions offered an opportunity to advance after the first halt, and I took advantage of it.

3. That the left being exposed by this advance of the line beyond the rest of the brigade, it was proper and necessary to re-enforce it by two companies.

4. That the two companies first deployed could not have reached the stone fort.

5. That the three companies added to the firing line gave it the power to reach the fort.

6. That the advance beyond the rest of the brigade was a bold and, without support, dangerous movement, but that the result justified the act. Had it failed I would have been held responsible.

7. That I saw at each stage of the battle what ought to be done, and did it. Results show that it was done at the right moment.

8. That the Twenty-fifth Infantry caused the surrender of the stone fort.

I desire to repeat that it is with great reluctance that I make so much of this report as relates to myself, and nothing but a sense of duty would impel me to do it.

Very respectfully,

A. S. DAGGETT,

Lieutenant-Colonel, Twenty-fifth Infantry, Commanding.

LOSSES OF THE TWENTY-FIFTH INFANTRY.

Killed.—Lieutenant H. L. McCorkle, Company G; Private Albert Strother, H; Private John W. Steele, D; Corporal Benj. Cousins, H; Private John B. Phelps, D; Private French Payne, B; Private Aaron Leftwich, G; Private Tom Howe, D.

Wounded.—Company A: Private William H. Clarke, Sergeant Stephen A. Browne. Company B: Private Tom Brown. Company C: Lieutenant John S. Murdock, Private Joseph L. Johnson, Private Samuel W. Harley, Private John A. Boyd. Company D; Captain Eaton A. Edwards, Sergeant Hayden Richards, Private Robert Goodwin. Company E: Lieutenant H. L. Kinnison, Private James Howard, Private John Saddler, Private David C. Gillam, Private Hugh Swann. Company F: First Sergeant Frank Coleman. Company G: Corporal James O. Hunter, Private Henry Brightwell, Private David Buckner, Private Alvin Daniels, Private Boney Douglas, Private George P. Cooper, Private John Thomas, Corporal Gov. Staton, Private Eugene Jones. Company H: Private James Bevill, Private Henry Gilbert.

Wounded July 2.—Private Elwood A. Forman, H; Private Smith, D; Private William Lafayette, F.

COMPLIMENTARY ORDER.

Headquarters 25th Infantry,
Near Santiago de Cuba, August 11, 1898.

General Orders No. 19.

The regimental commander congratulates the regiment on the prospect of its speedy return to the United States.

Gathered from three different stations, many of you strangers to each other, you assembled as a regiment for the first time in more than twenty-eight years on May 7, 1898, at Tampa, Florida. There you endeavored to solidify and prepare yourselves, as far as the oppressive weather would permit, for the work that appeared to be before you; but, who could have fortold the severity of that work?

You endured the severe hardships of a long sea voyage, which no one who has not experienced it can appreciate. You then disembarked, amidst dangerous surroundings; and on landing were for the first time on hostile ground. You marched, under a tropical sun, carrying blanket-roll, three days' rations, and one hundred rounds of ammunition, through rain and mud, part of the time at night, sleeping on the wet ground without shelter, living part of the time on scant rations, even, of bacon, hard bread and coffee, until on July 1 you arrived at El Caney. Here you took the battle formation and advanced to the stone fort, more like veterans than troops who had never been under fire. You again marched, day and night, halting only to dig four lines of intrenchments, the last being the nearest point to the enemy reached by any organization, when, still holding your rifles, within these intrenchments, notice was received that Santiago and the Spanish army had surrendered.

But commendable as the record cited may be, the brightest hours of your lives were on the afternoon of July 1. Formed in battle array, you advanced to the stone fort against volleys therefrom, and rifle-pits in front, and against a galling fire from block-houses, the church tower and the village on your left. You continued to advance, skilfully and bravely directed by the officers in immediate command, halting and delivering such a cool and well-directed fire that the enemy was compelled to wave the white flag in token of surrender.

Seldom have troops been called upon to face a severer fire, and never have they acquitted themselves better.

The regimental reserve was called upon to try its nerve, by lying quiet under a galling fire, without the privilege of returning it, where men were killed and wounded. This is a test of nerve which the firing line cannot realize, and requires the highest qualities of bravery and endurance.

You may well return to the United States proud of your accomplishments; and if any one asks you what you have done, point him to El Caney.

But in the midst of the joy of going home, we mourn the loss of those we leave behind. The genial, generous-hearted McCorkle fell at his post of duty, bravely directing his men in the advance on the stone fort. He died as the soldier dies, and received a soldier's burial. He was beloved by all who knew him, and his name will always be fondly remembered by his regiment—especially by those who participated in the Santiago campaign. The officers of the regiment will wear the prescribed badge of mourning for Lieutenant McCorkle for thirty days. And Corporal Benjamin Cousins, Privates Payne, Lewis, Strother, Taliaferro, Phelps, Howell, Steel and Leftwitch, sacrificed their lives on their country's altar. Being of a race which only thirty-five years ago emerged through a long and bloody war, from a condition of servitude, they in turn engaged in a war which was officially announced to be in the interest of humanity and gave all they had—their lives—that the oppressed might be free, and enjoy the blessings of liberty guaranteed by a stable government. They also died like true soldiers and received a soldier's burial.

By order of Lieutenant-Colonel Daggett.

M. D. CRONIN,

First Lieutenant and Adjutant, 25th Infantry.

MAJOR GENERAL AARON S. DAGGETT.

General Aaron S. Daggett is a native of Maine, born at Greene Corner, in that State, June 14, 1837. He is descended from a paternal ancestry which can be traced, with an honorable record, as far back as 1100 A. D. His mother was Dorcas C., daughter of Simon Dearborn, a collateral descendant of General Henry Dearborn. His more immediate ancestors came from Old to New England about 1630, and both his grandparents served in the Continental Army during the Revolutionary War. He was educated in his native town, at Monmouth Academy, Maine Wesleyan Seminary and Bates College. At the outbreak of the Civil War he enlisted as a private, April 27th, 1861, in the 5th Maine Infantry; was appointed second lieutenant May 1, and promoted first lieutenant May 24, 1861. He commanded his company at the first Bull Run battle, and was promoted captain August 14, 1861.

From the first engagement of the regiment to the end of its three years' memorable service, Captain Daggett proved a faithful and gallant soldier. He was promoted major, January 8th, 1863; on January 18th, 1865, was commissioned lieutenant-colonel of the 5th Regiment, United States Veteran Volunteers, Hancock Corps, and was brevetted colonel and brigadier-gentral of volunteers, March 13, 1865, for "gallant and meritorious services during the war." He also received the brevets of major in the United States Army for "gallant and meritorious services at the battle of Rappahannock Station, Va.," November 7, 1863, and lieutenant-colonel for "gallant and meritorious services in the battle of the Wilderness, Va." Immediately after the battle of Rappahannock Station, the captured trophies, flags, cannons, etc., were escorted, by those who had been most conspic-

Lieutenant-Colonel A. S. Daggett.

uous in the action, to General Meade's headquarters, Colonel Daggett being in command of the battalion of his brigade. General Upton to whom he owed this distinction, wrote of him as follows:

"In the assault at Rappahannock Station, Colonel Daggett's regiment captured over five hundred prisoners. In the assault at Spottsylvania Court House, May 10, his regiment lost six out of seven captains, the seventh being killed on the 12th of May, at the "angle," or the point where the tree was shot down by musketry, on which ground the regiment fought from 9.30 A. M. to 5.30 P. M., when it was relieved. On all these occasions Colonel Daggett was under my immediate command, and fought with distinguished bravery.

"Throughout his military career in the Army of the Potomac, he maintained the character of a good soldier and an upright man, and his promotion would be commended by all those who desire to see courage rewarded."

General Upton also wrote to the Governor of Maine as follows:

"I would respectfully recommend to Your Excellency, Major A. S. Daggett, formerly 5th Maine Volunteers, as an officer highly qualified to command a regiment. Major Daggett served his full term in this brigade with honor both to himself and State, and won for himself the reputation of being a brave, reliable and efficient officer. His promotion to a colonelcy would be a great benefit to the service, while the honor of his State could scarcely be entrusted to safer hands."

He was subsequently recommended for promotion by Generals Meade, Hancock, Wright and D. A. Russell. He was in every battle and campaign in which the Sixth Corps, Army of the Potomac, was engaged, from the first Bull Run to Petersburg, and was twice slightly wounded. On July 28, 1866, without his knowledge or solicitation, he was appointed a captain in the U. S. Regular Army, on recommendation of General Grant, and has since been promoted colonel in this service. During his subsequent career he has won the reputation of being a fine tactician and of being thoroughly versed in military law, as is indicated by Major Hancock's commendatory words in 1878:

"I look upon him as by far the best tactician in the regiment, and as for a thorough, clear knowledge of tactics his superior is not in the army. As regards military and civil law, I know of no one so well informed."

His ability and soldierly qualities have also been highly commended by General Crook, Colonel Hughes—Inspector-General in 1891—and Colonel ———, Inspector-General in 1892.

Not only as a soldier, but in many other ways, has General Daggett distinguished himself. As a public speaker the following was said of him by the Rev. S. S. Cummings, of Boston:

"It was my privilege and pleasure to listen to an address delivered by General A. S. Daggett on Memorial Day of 1891. I had anticipated something able and instructive, but it far exceeded my fondest expectations. The address was dignified, yet affable, delivered in choice language without manuscript, instructive and impressive, and highly appreciated by an intelligent audience."

General Daggett is noted for his courteous and genial manner, and his sterling integrity of character. He is a member of the Presbyterian church.

War Department, Inspector-General's Office,
Washington, Jan. 6th, 1899.

To the Adjutant-General, U. S. A., Washington, D. C.

Sir:—I desire to recommend to your favorable consideration and for advancement in case of the reorganization of the Regular Army, Lieutenant-Colonel A. S. Daggett, 25th U. S. Infantry.

I have known Colonel Daggett for a long time; he served in the War of the Rebellion with the 5th Maine Volunteers and acquitted himself with much honor; he served in Cuba in the war with Spain, commanding the 25th U. S. Infantry, and was conspicuous for gallantry at the battle of El Caney. He is an officer of the highest character, intelligent, courageous and energetic.

I sincerely trust that he may receive all the consideration he deserves.

Very respectfully,

(Sd) H. W. LAWTON,
Major-General, U. S. V.

A true copy:
M. D. CRONIN,
First Lieutenant and Adjutant 25th Infantry.
Headquarters Department of the East,
Governor's Island, New York City,
December 29, 1898.

Honorable R. A. Alger, Secretary of War, Washington, D. C.

Sir:—I recommend to the favorable consideration of the Secretary of War for promotion to Brigadier-General, Colonel A. S. Daggett, 25th Infantry. This officer has an excellent war record; his service has been faithful since then, and in the recent Spanish-American war he distinguished himself by his good judgment and faithful attention to duty, as well as for gallant service in action. An appointment of this character will be very highly appreciated throughout the army as a recognition of faithful, meritorious and gallant service. From my observation of Colonel Daggett he is well qualified for the position.

Very respectfully,
(Sd) WM. R. SHAFTER,
Major-General, U. S. Volunteers.

M. D. CRONIN,

A true copy:

First Lieutenant and Adjutant 25th Infantry.

To this very brilliant official record it is necessary to add but a word personal. Colonel Daggett is a typical New Englander; tall, well-formed, nervous and sinewy, a centre of energy, making himself felt wherever he may be. Precise and forceful of speech, correct and sincere in manners, a safe counsellor and a loyal friend, his character approaches the ideal. Stern and commanding as an officer he is nevertheless tender and sympathetic. His very sensitiveness concerning the feelings of others embarrasses him in giving expression to his own feelings on seeing suffering, unless it should be urgent, but those who know him best know him to be just, humane and tender. No man could have taken more care than he did for his regiment in Cuba. Hating oppression and wrong with a vehemency suited to his intense nature, he nevertheless deplores war and bloodshed. The President of the United States never did a more worthy act than when he gave to

Lieutenant-Colonel A. S. Daggett of the Twenty-fifth Infantry his commission as Brigadier-General of Volunteers in recognition of his valor and skill at El Caney and of his general efficiency as an officer in our army.

TESTIMONIES CONCERNING THE WORK OF THE TWENTY-FIFTH INFANTRY BEFORE EL CANEY.

Headquarters First District, Southern Luzon,
El Deposito, P. I., April 20, 1900.

My Dear General Daggett:—Some time ago I received a letter from you asking me to make an official statement as to where and at what objective the energies and fire of the 25th Infantry were directed during the battle of El Caney, Cuba, July 1, 1898.

In reply I have the honor to officially state that about noon July 1, 1898, the regiment moved from the mango grove, near the Ducro House, toward a stone fort located on a hill, near the town of El Caney.

It arrived at about one of the afternoon at a point about eight hundred yards to the south and east of the fort; immediately deployed, and the First Battalion, under command of Captain Walter S. Scott, and of which I was adjutant, designated as the attacking line. Presently, after advancing a few yards, we were subjected to a galling fire from the stone fort, the trenches in its front and from a blockhouse on its right. The line steadily moved forward, directing its fire at the stone fort and the trenches surrounding it. When within about one hundred and fifty yards from the fort the line was halted, and several sharpshooters, directed by their company officers to fire at the loopholes. Finally, when the men had regained their wind, a rush was made, part of the line going through a cornfield. At the foot the line was again halted, and after a few moments' rest charged up the hill, and the fort surrendered.

I went to the fort and found a Spanish lieutenant and seven enlisted men whom I passed out and were taken charge of by an officer of the 12th Infantry. This was about 3.50 P. M.

NOTE.—Since the above was written, General Daggett served with great distinction in the Philippines and in China, and was retired as a brigadier-general—a hero of four wars. A bill is now before Congress to make him a major-general, an honor to which he is most justly entitled.

As soon as the line reached the top of the hill it was fired on from the town, which had before been masked by the hill; the fire was of course returned, and this was the first fire from the battalion directed at the town. About five o'clock firing had ceased, the battalion was assembled and marched away.

(Sd) H. W. FRENCH,
First Lieutenant, 17th Infantry (late Second Lieutenant 25th Infantry.

A true copy:
H. G. LEARNARD,
Capt. and Adj. 14th Infantry.

Manila, P. I., March 30, 1900.

I certify that in the action of El Caney, Cuba, July 1, 1898, the company I commanded, i. e., H, 25th Infantry, directed its fire almost exclusively on the stone fort and the trench a few yards from its base. That very little of this company's fire was directed on the town and none before the fort was carried.

(Sd) VERNON A. CALDWELL,
First Lieutenant, 25th Infantry.

A true copy:
H. G. LEARNARD,
Capt. and Adj. 14th Infantry.

Tayug, Luzon, Philippine Islands,
April 17th, 1900.

To Those in Military Authority.

Regarding the battle of El Caney, Cuba, July 1, 1898, I hereby certify:

1. From about 1.20 o'clock P. M. to the time of the capture of the town of El Caney, I was in command of two companies —C and G—forming part of the 25th U. S. Infantry firing line.

2. From about 2.55 o'clock P. M. to the time of the capture of the town, very nearly the entire 25th Infantry firing line was under my observation.

3. From about 2.55 o'clock P. M. to about 3.20, the time of the surrender of the stone fort to the east of the town, the fire of the entire 25th Infantry firing line within my sight was directed against the fort.

4. During this period of the battle the 25th Infantry firing line was about 150 yards from the stone fort.

5. From the time the firing line began firing—about 1

o'clock P. M.—to the time of the surrender of the stone fort—about 3.20 P. M.—the companies under my command and all others under my observation concentrated their fire on the fort.

6. About 3.20 P. M., I was standing about 150 yards from the stone fort, and I plainly and distinctly saw a Spaniard appear in the door of the fort, and, for two or three seconds, wave a white flag at the 25th Infantry firing line, and upon being shot down, another Spaniard picked up the flag and likewise waved it at the 25th Infantry firing line.

7. After the white flag had twice been presented to the 25th Infantry firing line, and after all fire from the stone fort had ceased, the firing line rushed forward, took up a position facing to their left—that is, facing the town—and began a vigorous fire on a small blockhouse and on the town.

Respectfully, JAMES A. MOSS,
First Lieutenant, 24th U. S. Infantry.

RECOLLECTIONS OF THE SANTIAGO CAMPAIGN, BY CAPTAIN R. H. R. LOUGHBOROUGH, 25TH U. S. INFANTRY.

The 25th U. S. Infantry left its stations in Montana on the 10th of April, 1898; six companies (B, C, D, E, F and H) went in camp at Chickamauga National Park; the other two companies (A and G) went to Key West, Fla.

On May 6th the six companies at the Chickamauga National Park moved by rail to Tampa, Fla., arriving the night of the 7th, where they were joined by the two companies from Key West. With the exception of three days in 1870, the regiment had never been together since its organization in 1869. It necessarily followed that many of the officers, as well as men, were strangers to each other.

Our camp at Tampa was fair; the ground is sandy and flat, but as the rainy season had not set in, it was dry and the health was good. Drills and parades were held daily (Sundays excepted), but on account of the intense heat the hours for it were limited to the early mornings and after sunset. The clothing of the men was the same they had worn in Montana, and did not add to their comfort. Supplies of all kinds (except rations) came by piecemeal, and we finally sailed for the tropics with the same clothing used in the Northwest.

At 6 o'clock P. M. June 6th the regiment received orders to

strike tents and be ready to move within an hour; the order was immediately complied with, though the necessary transportation to move the baggage did not report until the forenoon of the following day; it was not far from noon when the last of it left the camp for the railroad station, en route to Port Tampa, where we were to embark on transports for the seat of war.

As soon as the camp equipage was started, the regiment was formed and marched to West Tampa (about three miles), where we took a train for Port Tampa, distant nine miles. On arrival, the regiment boarded the steamer "Concho," one of the vessels to carry the expedition to its destination. The 4th U. S. Infantry had preceded us, and the next day a battalion of the 2d Massachusetts Volunteers was put on, but owing to the crowded condition of the ship, a few days later they were transferred to another vessel.

The "Concho" is a large ship, but without the comforts I have seen since then on the U. S. Army transports plying between San Francisco and Manila. The ships used were hastily fitted up for the occasion, and it could not be expected that they would be all that was required, but some of the appointments could and should have been better. After a tedious wait until June 14th, we sailed down Tampa Bay and out on the Gulf of Mexico, still in ignorance of our destination. The evening of the 15th the light at Dry Tortugas was seen to our right. June 16th, 17th and 18th our course was a little south of east, and part of the time the north coast of Cuba was visible. The weather (except the intense heat) was fine. On Sunday morning, June 18th, we entered the Windward Passage, and it seemed certain, from our course, that Santiago was our objective. Early the next morning the high mountains of Santiago de Cuba were in plain sight to our north. June 20th and 21st, remained off the coast; the sea was rough and the vessel rolled considerably, adding to the discomfort of every one, especially those subject to seasickness. During the evening of the 21st, orders were received to be ready to disembark the following morning. About 8 A. M. on the 22d our warships began shelling the coast, and two hours later the troops started in small boats from the transports to the shore. By evening most of the Second Division and part of the Cavalry Division were on Cuban soil. There was no opposition to our landing; I believe that a small force well handled could have made it very difficult, if, indeed, it could not have prevented it.

As soon as the regiment had landed it was marched out about four miles and bivouacked for the night. The country is rugged and covered with a dense tropical vegetation. A few "Cuban Patriots" had joined us and formed the extreme advance, saving us some disagreeable outpost duty. This was the only service that I know of them doing throughout the campaign, though they were always on hand ration day. Later developments showed that the service rendered was not so important, as any Spanish force had retired to a safe place, something our friends looked out for whenever there was any danger.

June 23d, the regiment started shortly after daylight towards the city of Santiago. About 9 o'clock there was a report that the enemy were in our front. The regiment was immediately formed for battle, and reconnoitering parties sent forward; after about thirty minutes' delay the supposed enemy proved to be the large leaves of some tropical trees being moved by the wind, giving them the appearance of persons in motion. Our route was over a narrow trail, through a dense wilderness; water was scarce and the heat was intense. About noon we arrived at Siboney, where we bivouacked for the night. Before daylight next morning the troops in our rear were heard passing on the trail by our camp. Shortly after daylight Captain Capron's battery of four guns passed, and the men lined up along the road and cheered lustily. About an hour later, musketry fire and the occasional discharge of a Hotchkiss gun could be plainly heard towards Santiago. About three-quarters of an hour later we received orders to march. By mistake, the wrong trail was taken, and after marching fourteen hours we returned to our camp of the previous night, all fagged out. A great many men of the brigade were overcome with heat during this long, tiresome and fruitless ramble. I cannot say how many of these were of the 25th Infantry, but in my own company (B) there was not a man out of the ranks when the camp was reached. (I have called the above-mentioned place "Siboney." There is probably some other name for it, as the Cubans have one for every hamlet. It is not far from Siboney, and not knowing the name, have called it Siboney.)

On the morning of the 25th we got rations from the transport and all enjoyed a hearty breakfast. At 1 P. M. we broke camp and marched to Sevilla, about six miles. Here we remained until the morning of the 27th, part of the regiment being out on picket duty. June 27th, the regiment marched three miles towards Santiago and bivouacked on the banks of a small creek. Bathing was forbidden, as the creek was the only water supply for the army. The troops remained at this place until the afternoon of June 30th. The camp was in the valley of the creek, the ground is low and flat, and with the heavy rainfall every one was uncomfortable. Rations had to be brought from Siboney over a trail and did not arrive regularly.

About 1 o'clock in the afternoon on the 30th, the officers of the regiment were assembled at headquarters and were notified that there would be an attack on the Spanish position the next morning. About 4 o'clock the regiment started for its position, arriving after 10 o'clock, having covered a distance of less than three miles. The route was over an excuse for a road, but was crowded with some of the troops of almost every organization of the army, causing numberless halts, but worse than all, breaking the much-needed rest of the troops. On one part of this route I heard men asknig, "What regiment is this?" and heard various responses, as follows: "The W. W. W.'s, the 1st Cavalry, the 4th Infantry, the 10th Cavalry," etc. Some one asked, "What are the W. W. W.'s?" and some one replied. "Wood's Weary Walkers." I do not know who is responsible for that condition of affairs. Had we had an enterprising enemy in our front, disaster certainly would have followed. Here were a number of organizations scattered along a narrow,, muddy trail, at the mercy of an active foe. All this was only three or four miles from the Spanish works. The men were cheerful, and few if any realized that there might be danger.

Most of the men were up and moving about before daylight the next morning. Shortly after, the regiment started in the direction of El Caney. At 9 A. M. we halted in a mango grove near the Ducureau mansion. Shortly before noon a mounted orderly appeared with a message for the brigade commander. A few minutes later the march towards El Caney was taken up. Heavy musketry fire had been heard in that direction since shortly before 7 o'clock. A march of little more than a mile and the regiment was formed for battle, Companies G and H in the firing line, C and D in support, the remaining four companes in reserve.

For two hours or perhaps more the firing was very heavy, especially during the second hour. Attention is called to report of Colonel A. S. Daggett, pages 387 and 388, "Report of the War Department, 1898, Vol. I," and endorsement on same by

Major-General A. R. Chaffee. He says: "This stone fort was practically in the possession of the 12th Infantry at about 2 P. M. July 1." I cannot reconcile this statement with the fact that between the hours named some of the heaviest firing was going on, which does not indicate that its defenders were ready to give up. Lord Wellington once said, "At the end of every campaign truth lies at the bottom of a deep well, and it often takes twenty years to get her out." This may not be an exception. About half-past 4 o'clock the firing ceased and El Caney was ours.

The dead were collected near a hedge and the regiment was formed in column of masses to pay a silent tribute of respect to our departed comrades.

The regiment then started for the mango grove where we had left our blanket rolls and haversacks. Just as we were starting, some men with canteens started for water (about a mile away), when orders were received to be ready to march in twenty minutes. A few rods took us back to the road leading to Santiago. We moved down the road about three-quarters of a mile and halted. Two hours later, the pack train arrived with ammunition and then another with rations. Before the latter were issued orders were issued to move at once to the rear. The regiment marched over the trail it had come on the day before, arriving at El Poso about 8 o'clock A. M. Here we took the road leading to Santiago. About 9 A. M. we passed under San Juan Hill and moved to our right. Our forces held the crest of the hill. In passing along the hill we were sheltered from the fire except a short space, where one or two men were slightly wounded. Arriving at the La Cruz house near the road leading from El Caney to Santiago about 3. 30 P. M. and bivouacked for the night. About 10 o'clock the troops on our left were attacked by the Spanish. The firing was very heavy for an hour, when it suddenly ceased, and we retired for the night. During this time we were under the hill and protected from the fire.

Next morning (Sunday, July 3d) desultory firing began at daylight. About 7 A. M. the regiment left the La Cruz house and moved across the Caney-Santiago road and formed line to the left and moved forward to a ridge overlooking the city. A number of shots fell about us, but no one was struck. Shortly after, we were in possession of the ridge and began intrenching. The firing was kept up and two men were wounded. About noon we were informed that a truce had been established and all work was stopped. This gave all a much-needed rest, though it proved to be of short duration, caused by a false alarm by Major Webb, the inspector of the division staff.

During the afternoon the regiment was moved to the foot of the ridge, leaving only the pickets on the crest. About 8.30 P. M. we were ordered to the picket line and began intrenching. The tall grass was wet from a drenching rain a few hours before. The ground, though wet, was hard, and slow progress was made, having only their bayonets for picks and their bare hands for shovels. All night this work went on. The men were tired and hungry (as rations had not come up that day), but worked faithfully. During this, and I will add, throughout the campaign, I never heard a murmur nor a complaint; even when almost all the men of the regiment were down with fever and bowel trouble they were cheerful and ready to do any duty they were called on for.

The morning of July 3d Cervera's fleet sailed down the bay. An officer rode by our part of the line about half-past 9 and informed us of it. A few minutes later we heard the roar of the big guns, though at the time I little thought of what was going on. In the afternoon we heard cheering on our line way to the left, and as the good news came along it was taken up, and soon the whole line was shouting.

On the morning of July 5th the non-combatants left Santiago by two roads, one passing through our line. It was a pitiful sight. During the forenoon of the 5th we moved about a mile to the right and began intrenching. This position was very near the Spanish line, and quite elaborate works were constructed. We remained in this position until the morning of the 11th, when the regiment was ordered to the right of the line, about three miles. Here we intrenched. About 1 P. M. a truce was announced.

At 9.15 P. M. a staff officer came to the regimental commander's tent and informed him that the regiment was to be on the line at 12 o'clock midnight, and as soon as the moon rose to advance through the jungle until fired on, when the line was to halt and intrench. The night was stormy and any moon there might have been was obscured by the clouds. We were up, however, standing until daylight in a drenching rain, for it was so dark that any movement was impossible. Our rest was broken, without accomplishing anything that I know or heard of.

However, the rain and storm were providential, for I will always believe if the movement had been started we should have met with disaster. The ground was broken, deep ravines and underbrush with wire fences running through it. I have never learned who was "the father" of this order, and possibly never will. He must be ashamed of it.

The afternoon of the 12th the regiment advanced several hundred yards to the front and dug more intrenchments. They were still on this work the afternoon of the 14th when it was announced that the Spanish army had agreed to surrender. This came none too soon, for our men were coming down with malarial fever. A few days later nearly half the regiment were on the sick list, and the balance could not have done much.

The regiment was moved the same afternoon to higher ground in rear of the trenches. Strong guards were kept to look out for our prisoners and to prevent "our allies," the Cubans, from going into the city.

On the morning of the 17th the formal surrender of the city and Spanish army took place. We were some distance away and did not see anything of the ceremony.

On July 25th the regiment was moved about a mile further back in the hills and made camp, our tents, etc., having been brought up from the transport. Medicines appeared very scarce, resulting in much suffering. The food supplied was totally unfit for our new surroundings, and I believe not a little of the sickness can be traced to this. Our last camp was as good as any to be found in that vicinity.

The regiment remained in camp until August 13th, when it embarked on the transport "Camanche" for Montauk Point, arriving on the 18th, and landed on the 23d.

B. H. R. LOUGHBOROUGH,
Captain, 25th Infantry.

CHAPTER VII.

SAN JUAN.

Cavalry Division: The Ninth and Tenth Regiments.

When Lawton's division swung off to the right to engage the enemy at El Caney, with the results described in the preceeding chapter, the divisions of Wheeler and Kent were ordered to proceed directly along the Santiago road toward San Juan. Within a mile from El Pozo, the point where they had bivouacked for the night of the 30th, the troops arrived at the Aguadores River, which crosses the road here within less than a mile from San Juan Heights. Wheeler's division headed the column, although that general was not commanding. He had been relieved on the afternoon of the 30th and did not resume command until about 4 o'clock on July 1,* long after the heights had been carried, although he was on the field shortly after 1 o'clock of that day.

The Dismounted Cavalry Division on the morning of July 1 presented 2,663 fighting men, including officers. The First Brigade, commanded by Colonel Carrol, had 50 officers and 1,054 men, in regiments as follows: Third Cavalry, 22 officers, 420 men; Sixth Cavalry, 16 officers, 427 men; Ninth Cavalry, 12 officers, 207 men, the Ninth having hardly one-half the strength of either of the other regiments of the brigade. The Second Brigade, commanded by General Wood, contained 1,559 persons, distributed as follows: Brigade staff, 9 officers, 14 men; First Cavalry, 21 officers, 501 men; Tenth

*Official Report of General Sumner.

Cavalry, 22 officers, 507 men; First Volunteer Cavalry (Rough Riders), 25 officers, 517 men.

Before the troops left El Poso, Grimes' battery had been put in position and had fired a few shots at a blockhouse on San Juan Hill, distance 2,600 yards. Using black powder, which created a cloud of smoke with every shot, the battery was readily located by the foe, and the shrapnel from their guns was soon bursting among our forces. The second shot from the Spaniards wounded four of the Rough Riders and two or three of the regulars, while a third killed and wounded several Cubans. As a matter of course there was a rapid movements of the troops from that immediate vicinity. The firing soon ceased, and the troops took up that general advance movement already noted.

It is no easy task to follow the movements of the Cavalry Division from the time it left El Poso that July morning until it finally entrenched itself for the night on San Juan Hills. As heretofore we will take the official reports first, and from them make up the itinerary and the movements of the battle that followed, as far as they will enable us to do so. General Sumner says the division proceeded toward Santiago, and when about three-fourths of a mile from El Poso was halted in a narrow road to await orders and remained there for nearly an hour, subject to the effects of heavy artillery fire from the enemy's battery. Major Wessells, of the Third Cavalry, says, while following the road toward Santiago that morn, "much delay ensued from some reason unknown to the undersigned," and that the First Brigade of the division arrived at San Juan ford about 10 o'clock. This creek was about five hundred yards farther toward Santiago than Aguadores River, and ran about parallel with San Juan Heights, from which it was about three-fourths of a mile distant.

The orders for which General Sumner had waited nearly an hour under fire had come and were "verbal instructions to move to the San Juan Creek and hold it." Reaching this creek his advance guard was met by the Spaniards who fired one volley and retreated to a position on a hill on Sumner's right front, about 1,200 yards distant. Crossing this creek with sufficient strength to hold it, Sumner was now ordered to move by the right flank and connect with Lawton's left. While his troops were in this massed condition prior to deploying to the right through a thick jungle, the balloon that was in use for purposes of reconnoitering, came up the road and exposed itself to the full view of the Spaniards upon the heights. They needed no further invitation to direct toward our forces their artillery, for which the balloon became a flying target. Many officers and men were wounded here by exploding shells and small arms' fire of the enemy (Sumner). Under this fire, however, the troops were deployed as ordered.

Colonel Wood, who had charge of the Second Brigade, of which the Rough Riders were the leading regiment, says this "regiment was directed to change direction to the right, and by moving up the creek to effect a junction with General Lawton's division, which was engaged at Caney, about one and a-half miles toward the right, but was supposed to be working toward our right flank. After proceeding in this direction about half a mile the effort to connect with General Lawton was given up." This movement to the right took place between ten and eleven o'clock, at which time Lawton's forces had made no impression upon El Caney, and he was far from making any movement which might be described as working toward the right flank of the Cavalry Division. Lawton was not found by that half-hour's search to the right; and it was evident that something must be done by these troops in front, and done quickly. The whole division was under fire, and the battle on the Spanish side was in actual progress. True our men were hidden away in the jungle that bordered the creek, but their position was known to the Spaniards, and leaves and boughs are no cover from shot and shell. They were receiving the fire of the enemy and making no reply whatever, save by the few ineffective shots from the far away battery on El Poso Hill.

Directly in front of the cavalry division was a little hill occupied by a Spanish force. This hill is called in General Wood's report East Hill, but in the literature of the battle it is usaully mentioned as Kettle Hill. The fire in part was coming from here. Colonel Wood gives another report of the morning's experience in which he says: "The brigade moved down the road toward Santiago in rear of the First Brigade, with instructions to deploy to the right after crossing the San Juan, and continue to extend to the right, reaching out toward General Lawton's left and holding ourselves in rear of the First Brigade as a support. On reaching the stream the First Volunteer Cavalry, which was in the lead, crossed the stream with comparatively slight loss and deployed to the right in good order, but at this time a captive balloon was led down the road in which the troops were massed, and finally anchored at the crossing of the stream. The approach and anchoring of this balloon served to indicate the line of approach of the troops and to locate the ford, and the result was a terrific converging of artillery and rifle fire on the ford, which resulted in severe loss of men. Under this fire the First United States Cavalry and the Tenth United States Cavalry crossed the stream and deployed to the right where they were placed in position in rear of the First Brigade. Two regiments of the Second Brigade, to wit., the First and the Tenth Regular Cavalry, were located in the rear of the First Brigade. The First Regular Cavalry had begun its day's work as support of Grimes' battery, but had later come forward and taken its place in the brigade time enough to join in the action that followed.

"After completing the deployment," says Sumner, "the command was so much committed to battle that it became necessary either to advance or else retreat under fire." The troops were already in battle, but were not fighting, and could not do so in their present position, simply because they could not see the enemy. "Lieutenant Miley, representing General Shafter, authorized an advance, which was ordered, Carroll's brigade taking the advance, reinforced on the right by Roosevelt's regiment, and supported by the First and Tenth Cavalry." (Sumner.) Colonel Wood says: "After remaining in this position for about an hour (meaning the position held by his brigade previous to the coming of the order to advance) the order to advance was given, and the brigade advanced in good order as possible, but more or less broken up by the masses of brush and heavy grass and cactus; passing through the line of the First Brigade, mingling with them and charging the hill in conjunction with these troops, as well as some few infantry who had extended to the right." It must be remembered that the First Brigade consisted wholly of regulars, the Third, Sixth and Ninth Cavalry, while the Second Brigade had that remarkable regiment, the Rough Riders. This fact may account for their breaking through the lines of the First Brigade. Major Wessells, who commanded the Third Cavalry in that fight, and was himself wounded at the close of the first charge, says his regiment became entangled with other regiments, but, nevertheless, was to the crest as soon as any. Of the advance of the whole division, General Sumner says: "The advance was made under heavy infantry fire, through open flat ground, cut up by wire fences, to the creek, distant about 600 yards. The advance was made in good order, the enemy's fire being returned only under favorable opportunities. In crossing the flat one officer and several men were killed and several officers and men wounded. Both sides of the creek were heavily wooded for about 200 yards. The creek was swollen, and the crossing through this space and the creek was made with great difficulty.

"After passing through the thick woods the ground was entirely open and fenced by wire. From this line it was necessary to storm the hill, upon the top of which is a house, loopholed for defense. The slope of the hill is very difficult, but the assault was made with great gallantry and with much loss to the enemy. In this assault Colonel Hamilton, Lieutenants Smith and Shipp were killed; Colonel Carroll, Lieutenants Thayer and Myer were wounded. A number of casualties occurred among the enlisted men." The heights were carried by the whole division.

Lieutenant-Colonel Baldwin's account of the part his regiment took in the assault upon San Juan is told about as follows: After the search for Lawton had been given up, the First and Tenth Cavalry were formed for attack on East Hill. "I was directed," he says, "to take a position to the right, behind the river bank, for protection. While moving to this position, and while there, the regiment suffered considerable loss. After an interval of twenty or thirty minutes I was directed to form line of battle in a partially open field facing toward the blockhouses and strong intrenchments to the north occupied by the enemy. Much difficulty was found on account of the dense undergrowth, crossed in several directions by wire fences. As a part of the cavalry division under General Sumner, the regiment was formed in two lines, the First Squadron under Major S. T. Norvell, consisting of Troops A, B, E and I, leading; the second line, under Major T. J. Wint, consisting of Troops C, F and G. Troop D having crossed farther down the river, attached itself to a command of infantry and moved with that command on the second blockhouse. The regiment advanced in this formation in a heavy converging fire from the enemy's position, proceeding but a short distance when the two lines were united into one. The advance was rapidly continued in an irregular line toward the blockhouses and intrenchments to the right front. During this advance the line passed some troops of the First Cavalry, which I think had previously been formed on our right. Several losses occurred before reaching the top of the hill, First Lieutenant William H. Smith being killed as he arrived on its crest. The enemy having retreated toward the northwest to the second and third blockhouses, new lines were formed and a rapid advance was made upon these new positions. The regiment assisted in capturing these works from the enemy, and with the exception of Troops C and I, which in the meantime had joined the First Volunteer Cavalry, then took up a position to the north of the second blockhouse, remaining there all night."

Major Norvell, who commanded the First Squadron of the Tenth Cavelry, which consisted of Troops A, B, E and I, gives the following account of the experiences of July 1st:

"The regiment took position in a wood, and here suffered considerable loss, due to the fact that the whole of the enemy's fire appeared to be directed to this point. In a short time we moved out of the wood by the right flank and then deployed to the left, being then directly in front of the enemy and one mile distant from his works, marked by three houses about half a mile from one another. The enemy was strongly entrenched in front of these houses. The line, consisting of the cavalry division, under direction of Brigadier-General Sumner, moved forward in double time, under a terrific fire of the enemy. We had a very heavy jungle to march through, beside the river (San Juan) to cross, and during our progress many men were killed and wounded. The troops became separated from one another, though the general line was pretty well preserved. The works of the enemy were carried in succession by the troops; and the Spaniards were steadily driven back toward the town to their last ditches. We now found ourselves about half a mile from the city, but the troops being by this time nearly exhausted, here intrenched themselves for the night under a heavy fire. By dark this line was occupied by all the troops engaged during the day."

The official reports of the troop commanders of the Tenth Cavalry bring out a few more particulars which serve to give us a more vivid conception of this moving line. The entire cavalry division advanced together, and notwithstanding the roughness of the ground, Major Norvell assures us the line was pretty well preserved. Troops A, B, E and I were in the First Squadron, which was in the lead; Troops C, F and G were in the second line; Troop D made its advance with the infantry off to the left. We have now a fair knowledge of the general movement of the whole regiment. Let us follow the fortunes of some of the Troops, and by that means get nearer to the work done by the individual soldier.

Troop A was on the right of the leading squadron as the regiment took its place in line on the left of the First Cavalry and moved against the Spanish blockhouses in the face of a heavy fire, making a rush forward without intermission. A portion of the right platoon, under Lieutenant Livermore, became separated in one of the thickets, and under instructions received personally from the brigade commander, who seems to have been everywhere where he was needed, continued up the slope toward his right and toward the first blockhouse. The remainder of the troop, commanded by Captain Beck and Lieutenant McCoy, moved in the same direction at first, but observing that on account of the shorter distance to the slope from that end of the line, a large number of troops were arriving there, Captain Beck swung his troop to the left and reached the summit of the hill between the second and third blockhouses, and on arriving received a message by an aid of the brigade commander to hold the ridge. Just then Lieut. Livermore arrived, having come by way of Blockhouse No. 1. The troop now being together, held the crest for an hour. At times the fire of the enemy was so severe and Captain Beck's force so small that there was great danger that he would be compelled to abandon the position, but fortunately at the most critical juncture Lieutenant Lyon of the Twenty-fourth Infantry came up with a few reinforcements, and Lieutenant Hughes of the Tenth Cavalry with a Hotchkiss gun. Lieutenant Lyon formed his troops to the left of the gun, Troop A of the Tenth Cavalry being on the right. With this force the position was held until other troops arrived. Soon after, the squadron was reformed and the men entrenched themselves under fire. Troop B was next to Troop A and advanced as skirmishers by rushes and double time, but soon found its front blocked by other troops. Troop I advanced in two sections, the left being commanded by Lieutenant Miller, joined in the attack on the right of the enemy's position; the right commanded by Lieutenant Fleming, advanced on trenches between two blockhouses, and in so doing caught up with the rest of the troop. The first half of the troop, after attacking the blockhouse on right of the enemy's position then crossed the valley and attacked the blockhouse on the left of enemy's position, and then moved forward with the First Regular Cavalry and First Volunteer Cavalry, until the troop assembled as a whole. When it reached the place of intrenchment there were altogether about one hundred men at that point of the ridge, consisting of men from the Tenth Cavalry and of the Rough Riders. It is claimed by Lieutenant Anderson, who commanded Troop C, and who made his way to the front on the right of the line, that after coming up on the second hill and joining his troop to the left of Troop I, Colonel Roosevelt and part of his regiment joined on the right of the Tenth, and that he reported to him, placing C Troop in his command. Before this time Lieutenant Anderson had reported to Captain Jones, of Troop F, while they were on Kettle Hill, and the Two troops, F and C, had been formed in skirmish line and moved against the second blockhouse. In this movement Troop C got separated from Captain Jones, and Anderson, with 18 men of his own troop and several from other organizations, moved forward until he connected with Troop I, as previously narrated. These troops, C and I, were reported by their Colonel as having joined the First Volunteer Cavalry. All of the troop commanders who were immediately with the men bear hearty testimony to their good conduct. Captain Jones, commanding Troop F, says: "I could only do justice to the troop by mentioning by name all who were engaged, not only for their bravery, but for their splendid discipline under the most demoralizing fire." Lieutenant Fleming, commanding Troop I, says: "The entire troop behaved with great gallantry. Private Elsie Jones particularly distinguished himself." Captain Beck, commanding Troop A, says: "The behaviour of the enlisted men was magnificent, paying studious attention to orders while on the firing line, and generally exhibiting an intrepidity which marks the first-class soldier." Lieutenant Hughes, who commanded the Hotchkiss gun detachment, mentions four men for conspicuous bravery and commends his entire detachment for "spirit, enterprise and good behavior."

The official story is that the entire cavalry division advanced under orders from General Sumner and that the heft of its first blow fell upon Kettle Hill, which was soon captured, and on the crest of this hill the troops which had ascended it made a temporary halt, reformed their lines somewhat and immediately advanced upon the second hill to the help of that part of the cavalry division which had swung to the left in the advance, and also to the help of the infantry who were coming against Fort San Juan at the same time. Meanwhile there was left upon Kettle Hill a sufficient garrison or force to prevent its being recaptured by the enemy. In the assault on Kettle Hill the brigade commander, Colonel Carroll, had been wounded, and Lieutenant-Colonel Hamilton of the Ninth Cavalry killed. Many troop officers also had been either killed or wounded and also in the rush forward through the jungle and high grass some troops had been separated from their officers, and yet it is remarkable that all were ready to move forward to the next assault.

The words of praise to the whole cavalry division contained in the following order, published at Camp Wikoff immediately after the arrival there of the troops, are claimed by both black and white cavalrymen alike:

Headquarters, Cavalry Division,
Camp Wikoff, L. I., September 7th, 1898.

To the Officers and Soldiers of the Cavalry Division, Army of Santiago.

The duties for which the troops comprising the Cavalry Division were brought together have been accomplished.

On June 14th we sailed from Tampa, Fla., to encounter in the sickly season the diseases of the tropical island of Cuba. and to face and attack the historic legions of Spain in positions chosen by them and which for years they had been strengthening by every contrivance and art known to the skillful military engineers of Europe.

On the 23d, one squadron each of the 1st and 10th Regular Cavalry and two squadrons of the 1st Volunteer Cavalry, in all 964 officers and men, landed on Cuban soil. These troops marched on foot fourteen miles, and, early on the morning of the 24th, attacked and defeated double their number of regular Spanish soldiers under the command of Lieutenant-General Linares. Eagerly and cheerfully you pushed onward, and on July 1st forded San Juan River and gallantly swept over San Juan Hill, driving the enemy from its crest. Without a moment's halt you formed, aligning the division upon the 1st Infantry Division under General Kent, and, together with these troops, you bravely charged and carried the formidable intrenchments of Fort San Juan. The entire force which fought and won this great victory was less than seven thousand men.

The astonished enemy, though still protected by the strong works to which he had made his retreat, was so stunned by your determined valor that his only thought was to devise the quickest means of saving himself from further battle. The great Spanish fleet hastily sought escape from the harbor and was destroyed by our matchless navy.

After seizing the fortifications of San Juan Ridge, you, in

the darkness of night, strongly intrenched the position your valor had won. Reinforced by Bates' Brigade on your left and Lawton's Division on your right, you continued the combat until the Spanish army of Santiago Province succumbed to the superb prowess and courage of American arms. Peace promptly followed, and you return to receive the plaudits of seventy millions of people.

The valor displayed by you was not without sacrifice. Eighteen per cent., or nearly one in five, of the Cavalry Division fell on the field either killed or wounded. We mourn the loss of these heroic dead, and a grateful country will always revere their memory.

Whatever may be my fate, wherever my steps may lead, my heart will always burn with increasing admiration for your courage in action, your fortitude under privation and your constant devotion to duty in its highest sense, whether in battle, in bivouac or upon the march.

JOSEPH WHEELER,
Major-General U. S. V., Commanding.

Aside from that part of the Tenth Cavalry who fought under General Wheeler and who are consequently included among those congratulated by the General Order just quoted, Troop M of that regiment, under command of Lieutenant C. P. Johnson, performed an important part in the war. The troop consisted of 50 men and left Port Tampa June 21 on board the steamship Florida, the steamship Fanita also making a part of the expedition. The troop was mounted and was accompanied by a pack train of 65 animals. Both ships were heavily loaded with clothing, ammunition and provision, and had on board besides Lieutenant Johnson's command, General Nunez and staff and 375 armed Cubans. The expedition sailed around the west end of the island and attempted a landing at a point chosen by General Nunez on June 29, but failed owing to the fact that the place chosen was well guarded by Spaniards, who fired upon the landing party. The expedition had with it a small gunboat, the Peoria, commanded by

Captain Ryan, and on the afternoon of June 30th an attack was made upon a blockhouse on the shore by the gunboat, and a small force of Cuban and American volunteers landed, but were repulsed with the loss of one killed, General Nunez's brother, and seven wounded. Two days later Lieutenant Johnson was able to land and immediately made connection with General Gomez, unloading his stores for the Cuban Army.

Lieutenant G. P. Ahearn, of the Twenty-fifth Infantry, who went on this expedition as a volunteer, rendered important service on the night after the attack on the blockhouse at Tayabacoa. As the attacking party met with repulse and escaped to the ship in the darkness, several of their wounded were left on shore. Several boats sent out to recover them had returned without the men, their crews fearing to go on shore after them. Lieutenant Ahearn volunteered to attempt the rescue of the men, and taking a water-logged boat, approached the shore noiselessly and succeeded in his undertaking. The crew accompanying Lieutenant Ahearn was made up of men from Troop M, Tenth Cavalry, and behaved so well that the four were given Medals of Honor for their marked gallantry. The action of Lieutenant Ahearn in this case was in keeping with his whole military career. He has ever manifested a fondness for exceptional service, and has never failed when opportunity occurred to display a noble gallantry on the side of humanity. Nothing appeals to him so commandingly as an individual needing rescue, and in such a cause he immediately rises to the hero's plane. The noble colored soldiers who won medals on that occasion were all privates and became heroes for humanity's sake. Their names deserve a place in this history outside the mere official table. They were

Dennis Bell, George H. Wanton, Fitz Lee and William H. Tompkins, and were the only colored soldiers who, at the time of this writing, have won Medals of Honor in the Spanish War. Others, however, may yet be given, as doubtless others are deserved. The heroic service performed by whole regiments, as in the case of the Twenty-fourth Infantry, should entitle every man in it to a medal of some form as a souvenir for his posterity.

Losses of the Ninth Cavalry in the battles of San Juan:

OFFICERS—Killed, Lieutenant-Colonel John M. Hamilton.

MEN—Killed, Trumpeter Lewis Fort, Private James Johnson.

OFFICERS—Wounded, Adjutant Winthrop S. Wood, Captain Charles W. Taylor.

MEN—Wounded. First Sergeant Charles W. Jefferson, Sergeant Adam Moore, Sergeant Henry F. Wall, Sergeant Thomas B. Craig, Corporal James W. Ervine, Corporal Horace T. Henry, Corporal John Mason, Burwell Bullock, Elijah Crippen, Edward Davis, Hoyle Ervin, James Gandy, Edward D. Nelson, Noah Prince, Thomas Sinclair, James R. Spear, Jr., Jacob Tull, William H. Turner, George Warren, Alfred Wilson.

Losses of the Tenth Cavalry during the battle of San Juan:

OFFICERS—Killed, First Lieutenant W. E. Shipp, First Lieutenant W. H. Smith.

MEN—Killed, John H. Smoot, Corporal W. F. Johnson, John H. Dodson, George Stroal, William H. Slaughter.

OFFICERS—Wounded, Major T. J. Wint Captain John Bigelow, Jr., Adjutant and First Lieutenant M. H. Barnum, First Lieutenant R. L. Livermore, First Lieutenant E. D. Anderson, Second Lieutenant F. R. McCoy, Second Lieutenant

H. C. Whitehead, Second Lieutenant T. A. Roberts, Second Lieutenant H. O. Willard.

Men—Wounded, First Sergeant A. Houston, First Sergeant Robert Milbrown, Q. M. Sergeant William Payne, Sergeant Smith Johnson, Sergeant Ed. Lane, Sergeant Walker Johnson, Sergeant George Dyers, Sergeant Willis Hatcher, Sergeant John L. Taylor, Sergeant Amos Elliston, Sergeant Frank Rankin, Sergeant E. S. Washington, Sergeant U. G. Gunter, Corporal J. G. Mitchell, Corporal Allen Jones, Corporal Marcellus Wright, Privates Lewis L. Anderson, John Arnold, Charles Arthur, John Brown, Frank D. Bennett, Wade Bledsoe, Hillary Brown, Thornton Burkley, John Brooks, W. H. Brown, Wm. A. Cooper, John Chinn, J. H. Campbell, Henry Fearn, Benjamin Franklin, Gilmore Givens, B. F. Gaskins, William Gregory, Luther D. Gould, Wiley, Hipsher, Thomas Hardy, Charles Hopkins, Richard James, Wesley Jones, Robert E. Lee, Sprague Lewis, Henry McCormack, Samuel T. Minor, Lewis Marshall, William Matthews, Houston Riddill, Charles Robinson, Frank Ridgeley, Fred. Shackley, Harry D. Sturgis, Peter Saunderson, John T. Taylor, William Tyler, Isom Taylor, John Watson, Benjamin West, Joseph Williams, Allen E. White, Nathan Wyatt.

Note.—"While we talked, and the soldiers filled their canteens and drank deep and long, like camels who, after days of travel through the land of 'thirst and emptiness,' have reached the green oasis and the desert spring, a black corporal of the 24th Infantry walked wearily up to the 'water hole.' He was muddy and bedraggled. He carried no cup or canteen, and stretched himself out over the stepping-stones in the stream, sipping up the water and the mud together out of the shallow pool. A white cavalryman ran toward him shouting, 'Hold on, bunkie; here's my cup!' The negro looked dazed a moment, and not a few of the spectators showed amazement, for such a thing had rarely if ever happened in the army before. 'Thank you,' said the black corporal. 'Well, we are all fighting under the same flag now.' And so he drank out of the white man's cup. I was glad to see that I was not the only man who had come to recognize the justice of certain Constitutional amendments, in the light of the gallant behaviour of the colored troops throughout the battle, and, indeed, the campaign. The fortune of war had, of course, something to do with it in presenting to the colored troops the oportunities for distinguished service, of which they invariably availed themselves to the fullest extent; but the confidence of the general officers in their superb gallantry, which the event proved to be not misplaced, added still more, and it is a fact that the services of no four white regiments can be compared with those rendered by the four colored regiments—the 9th and 1toh Cavalry, and the 24th and 25th Infantry. They were to the front at La Guasima, at Caney, and at San Juan, and what was the severest test of all, that came later, in the yellow-fever hospitals."—Bonsal.

CHAPTER VIII.

SAN JUAN (Continued).

Kent's Division: The Twenty-fourth Infantry; Forming Under Fire—A Gallant Charge.

Turning now to the centre and left of the American line we follow the advance of that division of infantry commanded by General Kent, and which met the brunt of Spanish resistance at San Juan. This division, known as the First Division, Fifth Army Corps, consisted of three brigades, composed as follows:

First Brigade, Brigadier-General Hawkins commanding, made up of the Sixth Infantry, the Sixteenth Infantry, and the Seventy-first New York Volunteers.

The Second Brigade, Colonel Pearson commanding, made up of the Second Infantry, the Tenth Infantry and the Twenty-first Infantry.

The Third Brigade, commanded by Colonel Wikoff, in which were the Ninth Infantry, the Thirteenth Infantry and the Twenty-fourth Infantry; in all 262 officers and 5,095 men. Thus, in the whole division there were eight regiments of regular infantry and one volunteer regiment, the Seventy-first New York.

Although our present purpose is to bring into view the special work of the Twenty-fourth Infantry, it will be necessary to embrace in our scope the work of the entire division, in order to lay before the reader the field upon which that particular regiment won such lasting credit. General Kent, who commanded the division, a most accomplished soldier, gives a lucid account of the whole assault as seen from his position, and of the work performed by his division, in his report, dated July 8, 1898.

When General Kent's division arrived in the neighborhood of the San Juan ford and found itself under fire and the trail so blocked by troops of the cavalry division, which had not yet deployed to the right, that direct progress toward the front was next to impossible, the welcome information was given by the balloon managers that a trail branched off to the left from the main trail, only a short distance back from the ford. This trail led to a ford some distance lower down the stream and nearly facing the works on the enemy's right. General Kent on learning of this outlet immediately hastened back to the forks and meeting the Seventy-first New York Regiment, the rear regiment of the First Brigade, he directed that regiment into this trail toward the ford. The regiment was to lead the way through this new trail and would consequently arrive at the front first on the left; but meeting the fire of the enemy, the First Battalion of the regiment apparently became panic stricken and recoiled upon the rest of the regiment; the regiment then lay down on the sides of the trail and in the bushes, thoroughly demoralized.

Wikoff's brigade was now coming up and it was directed upon the same trail. This brigade consisted of the Ninth, Thirteenth and Twenty-fourth. Colonel Wikoff was directed by General Kent to move his brigade across the creek by the trail (the left fork) and when reaching the opposite side of the creek to put the brigade in line on the left of the trail and begin the attack at once. In executing this order the entire brigade stumbled through and over hundreds of men of the

Seventy-first New York Regiment. When a volunteer regiment broke through the lines of the Ninth Cavalry from the rear, that regiment was in its place on the field in line of battle, with its morale perfect. It was under discipline and delivering its fire with regularity. It had an absolute right to its place. The Seventy-first was in no such attitude, and General Kent directed the advance through it in these words: "Tell the brigade to pay no attention to this sort of thing; it is highly irregular." The Ninth Cavalry's position was exactly *regular*; the position of the Seventh-first was to the eyes of General Kent "highly irregular."

The three regiments of this brigade were to take their positions on the left of the ford after crossing the stream, in the following order: On the extreme left the Twenty-fourth, next to it in the centre of the brigade, the Ninth, and on the right of the brigade the Thirteenth. In approaching the ford the Ninth and Twenty-fourth became mixed and crossed in the following order: First one battalion of the Ninth; then a battalion of the Twenty-fourth; then the second battalion of the Ninth, followed by the second battalion of the Twenty-fourth. The line was formed under fire, and while superintending its formation the brigade commander, Colonel Wikoff, came under observation and was killed; Lieutenant-Colonel Worth, who succeeded him, was seriously wounded within five minutes after having taking command, and Lieutenant-Colonel Liscum, who next assumed charge of the brigade, had hardly learned that he was in command before he, too, was disabled by a Spanish shot. By this time, however, the formation was about complete and the brigade ready to begin the advance.

Leaving Wikoff's brigade in line ready to begin the advance

we must now return in our narrative to the main ford, where the major portions of Hawkins' and Pearson's brigades are massed and follow the various regiments as they come to their places in the battle line preparing for the onslaught. After crossing the ford with the Sixth Infantry, pursuant to the orders given by Lieutenant Miley in the name of General Shafter, General Hawkins attempted to flank the enemy by a movement to the left, the Sixth Infantry leading and the Sixteenth intending to pass beyond it in its rear and join to its left. The Sixth in passing to its intended position passed to the left of the Sixth Cavalry, which held the left of the line of the cavalry division, which had crossed the ford and deployed to the right, reaching beyond the Spanish lines in that direction, or at least it was able to reach the extreme right of the enemy. The Sixth Infantry continued this line southward and it was to be farther extended by the Sixteenth. Before this disposition could be effected the fire of the enemy became so severe that an advance movement was started and the Sixth lined up facing the fort on the hill, with only one company and a half of the Sixteenth on its left.

While Hawkins' and Wikoff's brigades were preparing for the advance upon the enemy's works, Pearson's brigade was approaching the ford, hurrying to the support. The Twenty-first Regiment of this brigade was detached from the brigade and sent directly forward on the main trail with orders to re-enforce the firing line. This regiment crossed the San Juan River to the left of the main ford and rushed forward to support Hawkins' left. In the meantime the two other regiments of the brigade, the Second and Tenth, which had preceded the Twenty-first in their march from El Poso, had been deflected to the left by order of the division commander and were passing to the front over the trail previously taken by Wikoff's brigade, crossing the San Juan at the lower ford. The Tenth crossed in advance and formed in close order on the opposite side of the stream, its line facing northwest. It was soon after, however, put in battle formation and moved to the right until it connected with the Twenty-first. The Second Regiment crossed the ford in the rear of the Tenth, having been delayed considerably by the Seventh-first New York Volunteers, who still blocked the way between the forks and the lower ford. After crossing the ford the Second put itself in line on the left of the Tenth, the whole brigade being now in position to support the First and Third Brigades in their charge.

This movement of Colonel Pearson's brigade had not been made without hardship and loss. All of the regiments came under the enemy's fire before reaching the San Juan River and many men were killed or wounded while the regiments were gaining their positions. The movement was so well executed as to call forth from the division commander the following enconium: "I observed this movement from the Fort San Juan Hill. Colonel E. P. Pearson, Tenth Infantry, commanding the Second Brigade, and the officers and troops under his command deserve great credit for the soldierly manner in which this movement was executed."

Although we left Wikoff's brigade standing in line on the left of the lower ford, we must not imagine that it remained in that position until the above movement on the part of the Second Brigade had been accomplished. There was no standing still in the fierce fire to which the men of that brigade were at that time subjected—a fire which had already cut down in rapid succession three brigade commanders. The formation was no sooner completed than the rapid advance began. The Thirteenth Infantry holding the right of the brigade moved to the right and front, while the Ninth and Twenty-fourth moved almost directly to the front at first, thus partially gaining the flank of the enemy's position. The whole line moved with great rapidity across the open field and up the hill, so that when the Second and Tenth Infantry came to their position as support, the heroic Third Brigade was well up the heights. To the right of the Third Brigade the First Brigade, containing the gallant Sixth, under Colonel Egbert, and the Sixteenth, was advancing also, and the two brigades arrived at the fort almost simultaneously; so that the division commander in speaking of the capture says: "Credit is almost equally due the Sixth, Ninth, Thirteenth, Sixteenth and Twenty-fourth Regiments of Infantry." To the Third Brigade he gives the credit of turning the enemy's right.

Let us now examine more closely that sweep of the Third Brigade from the left of the lower ford to San Juan Hill, in order to trace more distinctly the pathway of honor made for itself by the Twenty-fourth. This regiment formed left front into line under fire and advanced over the flat in good order, and then reformed under shelter of the hill preparatory to the final charge upon the enemy's intrenchments. The experience of the companies in crossing the flat is told by the company commanders. One company under the orders of its captain formed line of skirmishers and advanced in good order at rapid gait, reaching the foot of the hill almost exhausted. This was about the experience of all, but this company is mentioned because it was the first company of the regiment to reach the top of the hill. In crossing the flat there was necessarily some mixing of companies and in some instances men were separated from their officers, but those who escaped the enemy's bullets made their way across that plain of fire and were ready to join in the charge up the hill where only brave men could go.

There was but a moment's pause for breath at the foot of the hill and the general charge all along the line began, the Sixth Infantry probably taking the initiative, although the gallant Colonel Egbert, of that regiment (since killed in the Philippines), makes no such claim. In his farewell official report of the Sixth he thus describes the final act:

"We were now unexpectedly re-enforced. Lieutenant Parker, made aware by the heavy fire from the hill that a conflict was going on in his front, opened fire with his Gatlings most effectively on the intrenchments, while from far down on my left I heard cheering and shouts, and saw coming up the slope towards us a multitude of skirmishers. As they drew nearer we distinguished the tall figure of General Hawkins, with his aide, Lieutenant Ord, Sixth Infantry, charging at the head of the skirmishers and waving their hats. When the charge came up nearly abreast of where the Sixth stood in the road I ordered the companies out through the gaps in the wire fence to join it, and they complied with the same alacrity and enthusiasm that they had displayed in entering this bloody field. The Gatlings redoubled their fierce grinding of bullets on the Spanish, despite which there still came a savage fire from the blockhouse and trenches. Here the gallant Captain Wetherell, Sixth Infantry, fell, shot through the forehead, at the head of his company, and I received a Mauser bullet through the left lung, which disabled me. But the blood of the troops was now up, and no loss of officers or men could stop them. They charged up the incline until, coming to a steep ridge near the top, they were brought to a stand by the hail of bullets from the Gatlings against the summit. As soon as this could be stopped by a signal, the mingled troops of the Sixth, Sixteenth, Thirteenth and Twenty-fourth swept up and over the hill and it was won."

From testimony gathered on the evening of the fight it was concluded that there were more men of the Twenty-fourth Infantry on the ridge in this first occupation than of any other regiment, but all of the regiments of the division had done admirably and the brave blacks of the Twenty-fourth won on that day a standing in arms with the bravest of the brave.

The Spaniards although driven from their first line, by no means gave up the fight; but retreating to a line of intrenchments about eight hundred yards in the rear they opened upon the new-comers a fire almost as hot as before, and the troops found it difficult to hold what they had gained. The supporting regiments were coming up and strengthening the line, the men meanwhile entrenching themselves under fire as rapidly as possible. The Thirteenth Infantry was immediately ordered off to the right to assist the cavalry division, especially the Rough Riders, who were said to be in danger of having their flank turned. Here it remained under fire all night.

The advance and charge of the Twenty-fourth made up only a part of the advance and charge of the Third Brigade; and this in turn was part of the attack and assault made by the whole infantry division; a movement also participated in at the same hour by the cavalry division; so that regarded as a whole, it was a mighty blow delivered on the enemy's right and centre by two-thirds of the American Army, and its effect was stunning, although its full weight had not been realized by the foe. The part sustained in the assault by each regiment may be estimated by the losses experienced by each in killed and wounded. Judged by this standard the brunt fell upon the Sixth, Sixteenth, Thirteenth and Twenty-fourth, all of which regiments lost heavily, considering the short time of the action.

The movement by which the Twenty-fourth reached its position on that memorable 1st of July has called forth especial mention by the regimental commander and by the acting Assistant Adjutant-General of the brigade; it was also noted immediately after the battle by all the newspaper writers as one of the striking occurrences of the day. The regiment on coming under fire marched about one mile by the left flank, and then formed left front into line on its leading company, Company G, commanded by Captain Brereton. The first man of the regiment to take position in the line was the First Sergeant of G Company, R. G. Woods. This company when reaching its position formed on left into line, under a severe fire in front and a fire in the rear; the other companies forming in the same manner, with more or less regularity, to its left. As soon as the line was formed the order was given to charge. The advance was made across an open meadow, during which several officers were wounded, among them the officers of Company F, the command of that company devolving upon its First Sergeant, William Rainey, who conducted the company successfully to the crest of the hill.

The description of the movement of Company D as given by Lieutenant Kerwin, who was placed in command of that company after its officers had been shot, is a very interesting document. Lieutenant Kerwin claims to have made his report from "close inquiries and from personal observation." According to this report the company was led across the San

Juan Creek by its Captain (Ducat), the Second Lieutenant of the company (Gurney) following it, and keeping the men well closed up. While crossing, the company encountered a terrific fire, and after advancing about ten yards beyond the stream went through a wire fence to the right, and advanced to an embankment about twenty yards from the right bank of the stream. Here Captain Ducat gave the order to advance to the attack and the whole company opened out in good order in line of skirmishers and moved rapidly across the open plain to the foot of San Juan Hill. In making this movement across the plain the line was under fire and the brave Lieutenant Gurney was killed, and First Sergeant Ellis, Corporal Keys and Privates Robinson and Johnson wounded. It was a race with death, but the company arrived at the base of the hill in good form, though well-nigh exhausted. After breathing a moment the men were ready to follow their intrepid commander, Captain Ducat, up the hill, and at twelve o'clock they gained the summit, being the first company of the regiment to reach the top of the hill. Just as they reached the crest the brave Ducat fell, shot through the hip, probably by a Spanish sharpshooter, thus depriving the company of its last commissioned officer, and leaving its first sergeant also disabled.

The commander of the regiment speaks of its doings in a very modest manner, but in a tone to give the reader confidence in what he says. He became temporarily separated from the regiment, but made his way to the crest of the hill in company with the Adjutant and there found a part of his command. He says a creditable number of the men of his regiment reached the top of the hill among the first to arrive there. The commander of the Second Battalion, Captain Wygant, crossed the meadow, or flat, some distance ahead of the battalion, but as the men subsequently charged up the hill, he was unable to keep up with them, so rapid was their gait. It was from this battalion that Captain Ducat's company broke away and charged on the right of the battalion, arriving, as has been said, first on the top of the hill. As the regiment arrived Captain Wygant, finding himself the ranking officer on the ground, assembled it and assigned each company its place. Captain Dodge, who commanded Company C in this assault, and who subsequently died in the yellow fever hospital at Siboney, mentions the fact that Captain Wygant led the advance in person, and says that in the charge across the open field the three companies, C, B and H, became so intermixed that it was impossible for the company commanders to distinguish their own men from those of the other companies, yet he says he had the names of twenty men of his own company who reached the trenches at Fort San Juan in that perilous rush on that fiery mid-day. The testimony of all the officers of the regiment is to the effect that the men behaved splendidly, and eight of them have been given Certificates of Merit for gallantry in the action of July 1.

The losses of the regiment in that advance were numerous, the killed, wounded and missing amounted to 96, which number was swelled to 104 during the next two days. So many men falling in so short a time while advancing in open order tells how severe was the fire they were facing and serves to modify the opinion which was so often expressed about the time the war broke out, to the effect that the Spanish soldiers were wanting both in skill and bravery. They contradicted this both at El Caney and at San Juan. In the latter conflict they held their ground until the last moment and inflicted a loss upon their assailants equal to the number engaged in the defence of the heights. Since July 1, 1898, expatiation on the cowardice and lack of skill of the Spanish soldier has ceased to be a profitable literary occupation. Too many journalists and correspondents were permitted to witness the work of Spanish sharpshooters, and to see their obstinate resistance to the advance of our troops, to allow comments upon the inefficiency of the Spanish Army to pass unnoticed. Our army from the beginning was well impressed with the character of the foe and nerved itself accordingly. The bravery of our own soldiers was fully recognized by the men who surrendered to our army and who were capable of appreciating it, because they themselves were not wanting in the same qualities.

---

*"The intrenchments of San Juan were defended by two companies of Spanish infantry, numbering about two hundred and fifty to three hundred men. At about 11 o'clock in the morning reinforcements were sent to them, bringing the number up to about seven hundred and fifty men. There were two pieces of mountain artillery on these hills, the rest of the artillery fire against our troops on that day being from batteries close to the city."—In Cuba with Shafter (Miley), page 117.

CHAPTER IX.

THE SURRENDER, AND AFTERWARDS.

In the Trenches—The Twenty-fourth in the Fever Camp—Are Negro Soldiers Immune?—Camp Wikoff.

After the battle of El Caney the Twenty-fifth Infantry started for the mango grove, where the blanket rolls and haversacks had been left in the morning, and on its way passed the Second Massachusetts Volunteers standing by the roadside. This regiment had seen the charge of the Twenty-fifth up the hillside, and they now manifested their appreciation of the gallantry of the black regulars in an ovation of applause and cheers. This was the foundation for Sergeant Harris' reply when on another occasion seeing the manifest kind feelings of this regiment to the Twenty-fifth, I remarked: "Those men think you are soldiers." "They know we are soldiers," replied the sergeant. The regiment bivouacked in the main road leading from El Caney to Santiago, but sleep was out of the question. What with the passing of packtrains and artillery, and the issuing of rations and ammunition, the first half of the night gave no time for rest; and shortly after 12 o'clock, apprehensions of a Spanish attack put every one on the alert. At 3.30 the march to the rear was commenced and the entire division passed around by El Poso and advanced to the front by the Aguadores road, finally reaching a position on Wheeler's right about noon, July 2.

Subsequently the line of investment was extended to the right, the Cuban forces under General Garcia holding the extreme right connecting with the water front on that side of the city. Next to them came Ludlow's McKibben's and Chaffee's forces. In McKibben's brigade was the Twenty-fifth, which dug its last trench on Cuban soil on July 14th, on the railroad running out from Santiago to the northwest. This intrenchment was the nearest to the city made by any American organization, and in this the regiment remained until the surrender.

The Twenty-fourth remained entrenched over to the left, in General Kent's division, lying to the right of the 21st. This regiment (24th) had won great credit in its advance upon the enemy, but it was to win still greater in the field of humanity. Capt. Leavel, who commanded Company A, said: "It would be hard to particularize in reporting upon the men of the company. All—non-commissioned officers, privates, even newly joined recruits—showed a desire to do their duty, yea, more than their duty, which would have done credit to seasoned veterans. Too much cannot be said of their courage, willingness and endurance." Captain Wygant, who commanded the Second Battalion of the regiment, says: "The gallantry and bearing shown by the officers and soldiers of the regiment under this trying ordeal was such that it has every reason to be proud of its record. The losses of the regiment, which are shown by the official records, show the fire they were subjected to. The casualties were greater among the officers than the men, which is accounted for by the fact that the enemy had posted in the trees sharpshooters, whose principal business was to pick them off." There is no countenance given in official literature to the absurd notion maintained by some, that it was necessary for the officers of black troops to expose themselves unusually in order to lead their troops, and that this fact accounts for excessive losses among them. The fact is that the regular officer's code is such that he is compelled to occupy the place in battle assigned him in the tactics, and no matter how great his cowardice of heart may be, he must go forward until ordered to halt. The penalty of cowardice is something to be dreaded above wounds or even death by some natures. "Colored troops are brave men when led by white officers." (?) As a matter of fact there is very little leading of any sort by officers in battle. The officer's place is in the rear of the firing line, directing, not leading, and it is his right and duty to save his own life if possible, and that of every man in his command, even while seeking to destroy the enemy, in obedience to orders. The record of the Twenty-fourth for bravery was established beyond question when it swept across that open flat and up San Juan Hill on that hot mid-day of July 1st, 1898.

After lying in the trenches until July 15th, the news reached the camp of the Twenty-fourth that yellow fever had broken out in the army, and that a large hospital and pest-house had been established at Siboney. About 4 o'clock that day an order came to the commanding officer of the regiment directing him to proceed with his regiment to Siboney and report to the medical officer there. The regiment started on its march at 5.30, numbering at that time 8 companies, containing 15 officers and 456 men. Marching on in the night, going through thickets and across streams, the men were heard singing a fine old hymn:

> When through the deep waters I call thee to go,
> The rivers of woe shall not thee o'erflow;
> For I will be with thee, thy troubles to bless,
> And sanctify to thee they deepest distress.

In view of what was before them, the words were very appropriate. They arrived on the hill at Siboney at 3.30 on the morning of July 16th.

Without discussing the graphic story told by correspondents of the highest respectability describing the regiment as volunteering, to a man, to nurse the sick and dying at Siboney, we will rather follow the official records of their doings in that fever-stricken place. On arriving at Siboney on the morning of July 16, Sunday, Major Markely, then in command of the regiment, met Colonel Greenleaf of the Medical Department, and informed him that the Twenty-fourth Infantry was on the ground. Colonel Greenleaf was just leaving the post, but Major La Garde, his successor, manifested his great pleasure in seeing this form of assistance arrive. Such a scene of misery presented itself to Major Markely's eyes that he, soldier as he was, was greatly affected, and assured Major La Garde that he was prepared personally to sink every other consideration and devote himself to giving what assistance he could in caring for the sick, and that he believed his whole regiment would feel as he did when they came to see the situation. In this he was not mistaken. The officers and men of the Twenty-fourth Infantry did give themselves up to the care of the sick and dying, furnishing all help in their power until their own health and strength gave way, in some instances laying down even their lives in this noble work.

On the day of arrival seventy men were called for to nurse yellow fever patients and do other work about the hospital. More than this number immediately volunteered to enter upon a service which they could well believe meant death to some of them. The camp was so crowded and filthy that the work of cleaning it was begun at once by the men of the Twenty-fourth, and day by day they labored as their strength would permit, in policing the camp, cooking the food for themselves and for the hospital, unloading supplies, taking down and removing tents, and numberless other details of necessary labor. Despite all the care that could be taken under such conditions as were found at Siboney, the yellow fever soon overran the entire camp, and of the 16 officers of the regiment, 1 had died, 2 more were expected to die; 3 were dangerously ill, and 5 more or less so. Out of the whole sixteen there were but three really fit for duty, and often out of the whole regiment it would be impossible to get 12 men who could go on fatigue duty. Out of the 456 men who marched to Siboney only 24 escaped sickness, and on one day 241 were down. Those who would recover remained weak and unfit for labor. Silently, without murmuring, did these noble heroes, officers and men, stand at their post ministering to the necessities of their fellowman until the welcome news came that the regiment would be sent north and the hospital closed as soon as possible. On August 8 Major La Garde, more entitled to the honor of being classed among the heroes of Santiago than some whose opportunities of brilliant display were vastly superior, succumbed to the disease. The fact should be borne in mind that all of these men, officers, soldiers and surgeons, went upon this pest-house duty after the severe labors of assault of July 1-2, and the two weeks of terrible strain and exposure in the trenches before Santiago, and with the sick and wounded consequent upon these battles and labors—none were strong.

On July 16th, the day after the Twenty-fourth left the trenches, the surrender was made and on the next morning the final ceremonies of turning over Santiago to the American forces took place, and the soldiers were allowed to come out of their ditches and enter into more comfortable camps. The hardships of the period after the surrender were not much less than those experienced while in the lines.

On the 26th of August the Twenty-fourth Infantry, having obtained an honorable release from its perilous duty, marched out of Siboney with band playing and colors flying to go on board the transport for Montauk; but of the 456 men who marched into Siboney, only 198 were able to march out, directed by 9 out of the 15 officers that marched in with them. Altogether there were 11 officers and 289 men who went on board the transport, but all except the number first given were unable to take their places in the ranks. They went on board the steamer Nueces, and coming from an infected camp, no doubt great care was taken that the transport should arrive at its destination in a good condition. Although there was sickness on board, there were no deaths on the passage, and the Nueces arrived in port "one of the cleanest ships that came to that place." The official report states that the Nueces arrived at Montauk Point September 2, with 385 troops on board; 28 sick, no deaths on the voyage, and not infected. Worn out by the hard service the regiment remained a short time at Montauk and then returned to its former station, Fort Douglass, Utah, leaving its camp at Montauk in such a thoroughly creditable condition as to elicit official remark.

While the Twenty-fourth Infantry had without doubt the hardest service, after the surrender, of any of the colored regiments, the others were not slumbering at ease. Lying in the trenches almost constantly for two weeks, drenched with rains, scorched by the burning sun at times, and chilled by cool nights, subsisting on food not of the best and poorly cooked, cut off from news and kept in suspense, when the surrender finally came it found our army generally very greatly reduced in vital force. During the period following, from July 16th to about the same date in August the re-action fell with all its weight upon the troops, rendering them an easy prey to the climatic influences by which they were surrounded.* Pernicious malarial fever, bowel troubles and yellow fever were appearing in all the regiments; and the colored troops appeared as susceptible as their white comrades. The theory had been advanced that they were less susceptible to malarial fever, and in a certain sense this appears to be true; but the experience of our army in Cuba, as well as army statistics published before the Cuban War, do not bear out the popular view of the theory. The best that can be said from the experience of Cuba is to the effect that the blacks may be less liable to yellow fever and may more quickly rally from the effects of malarial fever. These conclusions are, however, by no means well established. The Twenty-fourth suffered excessively from fevers of both kinds, and in the judgment of the commanding officer of the regiment "effectually showed that colored soldiers were not more immune from Cuban fever than white," but we must remember that the service of the Twenty-fourth was exceptional. The Twenty-fifth Infantry lost but one man during the whole campaign from climatic disease, John A. Lewis, and it is believed that could he have received proper medical care his life would have been saved. Yet this regiment suffered severely from fever as did also the Ninth and Tenth Cavalry.

Arriving at Montauk* early the author had the opportunity to see the whole of the Fifth Army Corps disembark on its

*"After the surrender, dear Chaplain, the real trouble and difficulties began. Such a period, from July 14, 1898 to August 14, 1898, was never before known to human beings, I hope. The starving time was nothing

return from Cuba, and was so impressed with its forlorn appearance that he then wrote of it as coming home on stretchers. Pale, emaciated, weak and halting, they came, with 3,252 sick, and reporting 87 deaths on the voyage. But, as General Wheeler said in his report, "the great bulk of the troops that were at Santiago were by no means well." Never before had the people seen an army of stalwart men so suddenly transformed into an army of invalids. And yet while all the regiments arriving showed the effects of the hardships they had endured, the black regulars, excepting the Twenty-fourth Infantry, appeared to have slightly the advantage. The arrival of the Tenth Cavalry in "good condition" was an early cheering item in the stream of suffering and debility landing from the transports. Seeing all of the troops land and remaining at Camp Wikoff until its days were nearly numbered, the writer feels sure that the colored troops arrived from the front in as good condition as the best, and that they recuperated with marked comparative rapidity.

The chaplain of the Twenty-fifth Infantry, while en route to join his regiment at Montauk, thinking seriously over the condition of the men returning from such a hard experience, concluded that nothing would be more grateful to them than a reasonable supply of ripe fruit, fresh from the orchards and

---

to the fever time, where scores died per day. We were not permitted to starve; but had fever, and had it bad; semi-decayed beef, both from refrigerators and from cans. We had plenty of fever, but no clothing until very late; no medicine save a little quinine which was forced into you all the time, intermittent only with bad meat."—Extract from a soldier's letter.

While the Twenty-fifth Infantry was in camp at Chickamauga Park I was ordered to Xenia, Ohio, on recruiting duty, and on July 5, on seeing the reports of the wounded I asked officially to be ordered to my regiment. An order to that effect came about a month later, directing me to join my regiment by way of Tampa, Florida. Arriving in Tampa, my destination was changed by telegraph to Montauk Point, N. Y., whither I arrived a few days before the regiment did.

fields. He therefore sent a dispatch to the Daily Evening News, published in Bridgeton, N. J., asking the citizens of that community to contribute a carload of melons and fruits for the men of the Twenty-fifth, or for the whole camp, if they so wished. Subsequently mentioning the fact to the commanding officer of the regiment, Lieutenant-Colonel Daggett, he heartily commended the idea, believing that the fruit would be very beneficial. The good people of Bridgeton took hold of the matter heartily, and in a short time forwarded to the regiment more than four hundred of Jersey's finest watermelons, fresh from the vines. These were distributed judiciously and the health of the men began to improve forthwith. Soon five hundred more arrived, sent by a patriotic citizen of Philadelphia. These were also distributed. Ladies of Brooklyn forwarded peaches and vegetables, and supplies of all sorts now were coming in abundance. Our men improved so rapidly as to be the occasion of remark by correspondents of the press. They were spoken of as being apparently in good condition. While engaged in the work of supplying their physicial wants the chaplain was taken to task by a correspondent of Leslie's for being too much concerned in getting a carload of watermelons for his regiment, to go over to a graveyard and pray over the dead. The next day the chaplain made haste to go over to that particular graveyard to relieve the country from the crying shame that the correspondent had pointed out, only to find two men already there armed with prayer-books and one of them especially so fearful that he would not get a chance to read a prayer over a dead soldier, that the chaplain found it necessary to assure him that the opportunity to pray should not be taken from him; and thus another popular horror was found to be without reality.

The colored ladies of Brooklyn organized a Soldiers' Aid Society, and besides contributing in a general way, as already mentioned, also made and presented to the soldiers about four hundred home-made pies, which were most highly appreciated. They also prepared a tasty souvenir commemorative of the heroic work performed by the troops in Cuba, and expressive of high appreciation of the gallantry of the colored regiments. A beautiful stand of colors was also procured for the Twenty-fourth Infantry, which were subsequently presented to the regiment with appropriate ceremonies.

At the camp were three colored chaplains and one colored surgeon, serving with the Regular Army, and their presence was of great value in the way of accustoming the people at large to beholding colored men as commissioned officers. To none were more attention shown than to these colored men, and there was apparently no desire to infringe upon their rights. Occasionally a very petty social movement might be made by an insignificant, with a view of humiliating a Negro chaplain, but such efforts usually died without harm to those aimed at and apparently without special comfort to those who engineered them.

The following paragraphs, written while in camp at the time indicated in them, may serve a good purpose by their insertion here, showing as they do the reflections of the writer as well as in outlining the more important facts associated with that remarkable encampment:

CAMP WIKOFF AND ITS LESSONS.

Now that the days of this camp are drawing to a close it is profitable to recall its unique history and gather up some of the lessons it has taught us. Despite all the sensationalism, investigations, testings, experimentation, and general condemnation, the camp at Montauk accomplished what was intended, and was itself a humane and patriotic establishment. It is not for me to say whether a better site might not have been selected, or whether the camp might not have been better managed. I will take it for granted that improvement might have been made in both respects, but our concern is rather with what was, than with what "might have been."

To appreciate Camp Wikoff we must consider two things specially; first, its purpose, and secondly, the short time allowed to prepare it; and then go over the whole subject and properly estimate its extent and the amount of labor involved.

The intention of the camp was to afford a place where our troops, returning from Cuba, prostrated with climatic fever, and probably infected with yellow fever, might receive proper medical treatment and care, until the diseases were subdued. The site was selected with this in view, and the conditions were admirably suited to such a purpose. Completely isolated, on dry soil, with dry pure air, cool climate, away from mosquitoes, the camp seemed all that was desired for a great field hospital.

Here the sick could come and receive the best that nature had to bestow in the way of respite from the heat, and pure ocean breezes, and, taken altogether, the experiences of August and a good part of September, have justified the selection of Montauk. While prostrations were occurring elsewhere, the camp was cool and delightful most of the time.

As to the preparations, it must be remembered that the recall of the whole Army of Invasion from Cuba was made in response to a popular demand, and as a measure of humanity. Bring the army home! was the call, and, Bring it at once!

Compliments
of the
Patriotic Colored Women
of
BROOKLYN
N. Y.
As the
"Black St. Domingo Legion"
saved the Patriot Army in the
Siege of Savannah
1779
So our noble "Black Heroes" came
to the rescue, and the victory
was gloriously won in the
Battle of Santiago
1898

Such urgency naturally leaps ahead of minor preparations. The soldiers wanted to come; the people wanted them to come; hence the crowding of transports and the lack of comforts on the voyages; hence the lack of hospital accommodations when the troops began to arrive. Haste almost always brings about such things; but sometimes haste is imperative. This was the case in getting the army out of Cuba and into Camp at Montauk in August, '98. Haste was pushed to that point when omissions had to occur, and inconvenience and suffering resulted.

We must also remember the condition of the men who came to Montauk. About 4,000 were reported as sick before they left Cuba; but, roughly speaking, there were 10,000 sick men landing in Montauk. Those who were classed as well were, with rare exceptions, both mentally and physically incapable of high effort. It was an invalid army, with nearly one-half of its number seriously sick and suffering.

Ten thousand sick soldiers were never on our hands before, and the mighty problem was not realized until the transports began to emit their streams of weakness and walking death at Montauk. The preparation was altogether inadequate for such a mass of misery, and for a time all appeared confusion.

Then came severe, cruel, merciless criticisms; deserved in some cases no doubt, but certainly not everywhere. The faults, gaps, failures, were everywhere to be seen, and it was easy to see and to say what ought to have been done. But the situation at Camp Wikoff from August 15th to Sep. 15th needed more than censure; it needed help. The men who were working for the Government in both the medical and commissary departments needed assistance; the former in the way of nurses, and the latter in the way of appropriate food. The censure and exposure indulged in by the press may have contributed to direct the attention of the benevolently disposed to the conditions in the camp.

Then came the era of ample help; from Massachusetts; from New York, in a word, from all over the country. The Merchants' Relief Association poured in its thousands of dollars worth of supplies, bringing them to the camp and distributing them generously and wisely. The Women's Patriotic Relief, the Women's War Relief, the International Brotherhood League, and the powerful Red Cross Society, all poured in food and comforts for the sick thousands. Besides these great organizations there were also the spontaneous offerings of the people, many of them generously distributed by the Brooklyn Daily Eagle's active representatives. The tent of that journal was an excellent way-mark and a veritable house of the good shepherd for many a lost wanderer, as well as a place of comfort, cheer and rest. The work done was very valuable and highly appreciated.

To the medical department came the trained hand of the female nurse. No one who saw these calm-faced, white-hooded sisters, or the cheery cheeked, white capped nurses from the schools, could fail to see that they were in the right place. The sick soldier's lot was brightened greatly when the gentle female nurse came to his cot. Woman can never be robbed of her right to nurse. This is one of the lessons taught by the Hispano-American War.

This vast army has been handled. No yellow fever has been spread. The general health has been restored. The disabled are mostly housed in hospitals, and many of them are on the road to recovery. Some have died; some are on furlough, and many have gone to their homes.

The regulars are repairing to their stations quite invigorated, and greatly helped in many ways by the kind treatment they have received. Camp Wikoff was not a failure; but a great and successful object lesson, as well as a great summer school in nationalism. Here black, white and Indian soldiers fraternized; here Northerners and Southerners served under the same orders. Ten thousand soldiers and as many civilians daily attended the best school of its kind ever held in this country, striving to take home to their hearts the lessons that God *is* teaching the nations.

The Rev. Sylvester Malone thus sums up the message of the war to us in his letter to the committee to welcome Brooklyn's soldiers:

"This short war has done so much for America at home and abroad that we must take every soldier to our warmest affection and send him back to peaceful pursuits on the conviction that there is nothing higher in our American life than to have the privilege to cheer and gladden the marine and the soldier that have left to America her brightest and best page of a great history. This past war must kindle in our souls a love of all the brethren, black as well as white, Catholic as well as Protestant, having but one language, one nationality, and it is to be hoped, yet one religion."

These are true words, as full of patriotism as they are of fraternity, and these are the two special lessons taught at Montauk—a broad, earnest, practical fraternity, and a love of country before which the petty prejudices of race and section were compelled to yield ground.

THE YOUNG MEN'S CHRISTIAN ASSOCIATION IN CAMP WIKOFF.

The Young Men's Christian Association has done an excellent work in Camp Wikoff. Their tents have afforded facilities for profitable amusements, in the way of quiet games, thus bringing out the use of these games distinct from their abuse—gambling.

Their reading tables have also been well supplied with papers and magazines, religious and secular, generally very acceptable to the soldiers, as attested by the numbers that read them. But perhaps best of all, has been the provision made for the soldiers to write. Tables, pens, ink, paper and envelopes have been supplied in abundance. These were of great advantage to soldiers living in tents, and the work of the Association in this respect cannot be too highly commended.

The specially religious work of the Association as I have seen it, consists of three divisions: First, the meetings in their tents, held nightly and on Sundays. These have been vigorously carried on and well attended, the chaplains of the camp often rendering assistance. Secondly, I have noticed the Y. M. C. A. men visiting the sick in the hospitals and camps, giving the word of exhortation and help to the sick. Perhaps, however, in their work of private conversation with the well men, they have done as much real service for God as in either of the other two fields. They have made the acquaintance of many men and have won the respect of the camp. This I have numbered as the third division of their work—personal contact with the soldiers of the camp, at the same time keeping themselves "unspotted from the world."

B.

The 24th Infantry was ordered down to Siboney to do guard duty. When the regiment reached the yellow-fever hospital it was found to be in a deplorable condition. Men were dying there every hour for the lack of proper nursing. Major Markley, who had commanded the regiment since July 1st, when Colonel Liscum was wounded, drew his regiment up in line, and Dr. La Garde, in charge of the hospital, explained the needs of the suffering, at the same time clearly setting forth the danger to men who were not immune, of nursing and attending yellow-fever patients. Major Markley then said that any man who wished to volunteer to nurse in the yellow-fever hospital could step forward. The whole regiment stepped forward. Sixty men were selected from the volunteers to nurse, and within forty-eight hours forty-two of these brave fellows were down seriously ill with yellow or pernicious malarial fever. Again the regiment was drawn up in line, and again Major Markley said that nurses were needed, and that any man who wished to do so could volunteer. After the object lesson which the men had received in the last few days of the danger from contagion to which they would be exposed, it was now unnecessary for Dr. La Garde to again warn the brave blacks of the terrible contagion. When the request for volunteers to replace those who had already fallen in the performance of their dangerous and perfectly optional duty was made again, the regiment stepped forward as one man. When sent down from the trenches the regiment consisted of eight companies, averaging about forty men each. Of the officers and men who remained on duty the forty days spent in Siboney, only twenty-four escaped without serious illness, and of this handful not a few succumbed to fevers on the voyage home and after their arrival at Montauk.

As a result, thirty-six died and about forty were discharged from the regiment owing to disabilities resulting from sickness which began in the yellow-fever hospital.—Bonsal's Fight for Santiago.

CHAPTER X.

REVIEW AND REFLECTIONS.

Gallantry of the Black Regulars—Diary of Sergeant-Major E. L. Baker, Tenth Cavalry.

It is time now to sum up the work of the four regiments whose careers we have thus far followed, and to examine the grounds upon which the golden opinions they won in battle and siege are based. We have seen that in the first fight, that of Las Guasimas, on June 24th, the Tenth Cavalry, especially Troops I and B, both with their small arms and with the machine guns belonging to Troop B, did most effective work against the Spanish right, joining with the First Cavalry in overcoming that force which was rapidly destroying Roosevelt's Rough Riders. Nor should it be forgotten that in this first fight, Troop B, which did its full share, was commanded on the firing line by Sergeants John Buck and James Thompson. In the squad commanded by Sergeant Thompson several men of the First Regular Cavalry fought and it is claimed were highly pleased with him as squad commander.

While this was the firrst fight of the men of the Tenth Cavalry with the Spaniards, it was by no means their first experience under fire. From the time of the organization of the regiment in 1866 up to within a year of the war, the men had been engaged frequently in conflicts with Indians and marauders, often having men killed and wounded in their ranks. The fights were participated in by small numbers, and the casualties were not numerous, but there were opportunities for the acquirement of skill and the display of gallantry. Altogether the men of the regiment during their experience on the plains engaged in sixty-two battles and skirmishes. This training had transformed the older men of the regiment into veterans and enabled them to be cool and efficient in their first fight in Cuba.

Sergeant Buck, upon whom the command of Troop B chiefly fell after becoming separated from his Lieutenant in the battle at Guasimas, joined the regiment in 1880, and had already passed through eighteen years of the kind of service above described. He was at the time of the Cuban War in the prime of life, a magnificent horseman, an experienced scout, and a skilled packer. In 1880, when he joined the regiment, the troops were almost constantly in motion, marching that one year nearly seventy-seven thousand miles, his own troop covering twelve hundred and forty-two miles in one month. This troop with four others made a ride of sixty-five miles in less than twenty-one hours, arriving at their destination without the loss of a single horse. In 1893 he was mentioned by the commanding officer of Fort Missoula, Montana, for highly meritorious service, skill and energy displayed while in charge of pack train of an expedition across the Bitter Root Mountains, Idaho, during the most inclement weather, in quest of a party of gentlemen lost. (Letter of commanding officer, Fort Missoula, Montana, February 12, 1894.) Sergeant Buck has also won the silver medal for revolver shooting.

Sergeant James Thompson joined the regiment in 1888, and has passed the ten years in the one troop, and proved himself at Las Guasimas a soldier worthy his regiment.

The first battle gave the Tenth a reputation in a new field, corresponding to that which it had gained in the West, and this was not allowed to fade during its stay in Cuba. The fame of this first action spread rapidly through the army and inspired the other regiments of colored men with a desire to distinguish themselves on this new field of honor, and their readiness to be to the front and to take prominent part in all service was so marked that opportunity could not be withheld from them. As the army advanced toward Santiago these regiments became more and more the mark of observation by foreign military men who were present, and by the great throng of correspondents who were the eyes for the people of the civilized world. And hence, when the lines of assault were finally determined and the infantry and cavalry of our army deployed for its perilous attack upon the Spanish fortifications the black regiments were in their places, conspicuous by their vigor and enthusiasm. In them were enlisted men whose time of service had expired a few days before, but who had promptly re-enlisted. In at least two cases were men who served their full thirty years and could have retired with honor at the breaking out of the war. They preferred to share the fortunes of their comrades in arms, and it is a comfort to be able to record that the two spoken of came home from the fight without a wound and with health unimpaired. How many others there were in the same case in the army is not reported, but the supposition is that there were several such in both the white and colored regiments.

Recalling the scenes of that memorable first of July, 1898, we can see the Twenty-fifth Infantry advancing steadily on the stone fort at El Caney at one time entirely alone, meeting the fire of the fort even up to their last rush forward. Captain Loughborough, who commanded Company B, of that regiment, and although his company was in the reserve, was nevertheless under fire, says: "The hardest fighting of the Twenty-fifth was between two and four o'clock," at which time all the other troops of the attacking force, except Bates' brigade, were under cover and remaining stationary, the Twenty-fifth being the only organization that was advancing. The official reports give the positions of General Chaffee's brigade during the two hours between two o'clock and four of that afternoon as follows:

The Seventh was under partial cover and remained in its position "until about 4.30 p. m." The Seventeenth remained with its left joined to the right of the Seventh "until the battle was over." The Twelfth Infantry was in its shelter within 350 yards of the fort "until about 4 p. m." Ludlow's brigade was engaged with the town, hence only Miles' brigade, consisting of the Fourth and Twenty-fifth Infantry, was advancing upon the fort. The Fourth Infantry was soon checked in its advance, as General Daggett especially notes in his report, and the Twenty-fifth was thus thrust forward alone, excepting Bates' brigade, which was making its way up the right.

This conspicuous advance of the Twenty-fifth brought that regiment into the view of the world, and established for it a brilliant reputation for skill and courage. Arriving in the very jaws of the fort the sharpshooters and marksmen of that regiment poured such a deadly fire into the loopholes of the fort that they actually silenced it with their rifles. These men with the sterness of iron and the skill acquired by long and careful training, impressed their characteristics on the minds of all their beholders. Of the four hundred men who went on the field that morning very few were recruits, and many had passed over ten years in the service. When they "took the battle formation and advanced to the stone fort more like veterans than troops who had never been under fire," as their commander reports, they gave to the world a striking exhibition of the effect of military training. In each breast a spirit of bravery had been developed and their skill in the use of their arms did not for a moment forsake them. They advanced against volleys from the fort and rifle pits in front, and a galling fire from blockhouses, the church tower and the village on their left. Before a less severe fire than this, on that very day, a regiment of white volunteers had succumbed and was lying utterly demoralized by the roadside; before this same fire the Second Massachusetts Volunteers were forced to retire—in the face of it the Twenty-fifth advanced steadily to its goal.

Lieutenant Moss, who commanded Company H on the firing line on that day, has published an account in which he says: "The town was protected on the north by three blockhouses and the church; on the west by three blockhouses (and partially by the church); on the east by the stone fort, one blockhouse, the church, and three rifle pits; on the south and southeast by the stone fort, three blockhouses, one loop-holed house, the church and eight rifle pits. However, the Second Brigade was sent forward against the southeast of the town, thus being exposed to fire from fourteen sources, nearly all of which were in different planes, forming so many tiers of fire. The cover on the south and southeast of the town was no better than, if as good, as that on the other sides."

The cavalry regiments were no less conspicuous in their gallantry at San Juan than was the Twenty-fifth Infantry at El Caney. The brilliancy of that remarkable regiment, the Rough Riders, commanded on July 1st by Colonel Roosevelt, was so dazzling that it drew attention away from the ordinary regulars, yet the five regiments of regular cavalry did their duty as thoroughly on that day as did the regiment of volunteers.* In this body of cavalry troops, where courage was elevated to a degree infringing upon the romantic, the two black regiments took their places, and were fit to be associated in valor with that highly representative regiment. The Inspector-General turns aside from mere routine in his report long enough to say "the courage and conduct of the colored troops and First United States Volunteers seemed always up to the best." That these black troopers held no second place in valor is proven by their deeds, and from the testimony of all who observed their conduct, and that they with the other regulars were decidedly superior in skill was recognized by the volunteer Colonel himself. The Ninth Cavalry, although suffering considerably in that advance on East Hill, involved as it was, more or less, with Roosevelt's regiment, did not receive so large a share of public notice as its sister regiment. The strength of the Ninth was but little over one-half that of the Tenth, and its movements were so involved with those of the volunteers as to be somewhat obscured by them; the loss also of its commander just as the first position of the enemy fell into our hands, was a great misfortune to the regiment. The Ninth, however, was with the first that mounted the heights, and whatever praise is to be bestowed upon the Rough Riders in that assault is to be distributed in equal degree to the men of that regiment. Being in the leading brigade of the division this regiment had been firing

*"The Ninth and Tenth Cavalry regiments fought one on either side of mine at Santiago, and I wish no better men beside me in battle than these colored troops showed themselves to be. Later on, when I come to write of the campaign, I shall have much to say about them."—T. Roosevelt.

steadily upon the Spanish works before the charge was ordered, and when the movement began the men of the Ninth advanced so rapidly that they were among the first to reach the crest.

The Tenth Regiment, with its Hotchkiss guns, and its trained men, took its place in the line that morning to add if possible further lustre to the distinction already won. In crossing the flat, in climbing the heights, and in holding the ridge these brave men did all that could be expected of them. Roosevelt said: "The colored troops did as well as any soldiers could possibly do," meaning the colored men of the Ninth and Tenth Cavalry. To their officers he bestows a meed of praise well deserved, but not on the peculiar ground which he brings forward. He would have the reader believe that it has required special ability and effort to bring these colored men up to the condition of good soldiers and to induce them to do so well in battle; while the testimony of the officers themselves and the experience of more than a quarter of a century with colored professional troops give no countenance to any such theory. The voice of experience is that the colored man is specially apt as a soldier, and General Merritt declares him always brave in battle. The officers commanding colored troops at Santiago honored themselves in their reports of the battles by giving full credit to the men in the ranks, who by their resolute advance and their cool and accurate firing dislodged an intrenched foe and planted the flag of our Union where had floated the ensign of Spain.

That rushing line of dismounted cavalry, so ably directed by Sumner, did not get to its goal without loss. As it swept across the open to reach the heights, it faced a well-directed fire from the Spanish works, and men dropped from the ranks, wounded and dying. Of the officers directing that advance 35 fell either killed or wounded and 328 men. These numbers appear small when hastily scanned or when brought into comparison with the losses in battle during the Civil War, but if we take time to imagine 35 officers lying on the ground either killed or wounded and 328 men in the same condition, the carnage will not appear insignificant. Woe enough followed even that one short conflict. It must be observed also that the whole strength of this division was less than 3000 men, so that about one out of every eight had been struck by shot or shell.

Several enlisted men among the colored cavalry displayed high soldierly qualities in this assault, evidencing a willingness to assume the responsibility of command and the ability to lead. Color-Sergeant George Berry became conspicuous at once by his brilliant achievement of carrying the colors of two regiments, those of his own and of the Third Cavalry. The Color-Sergeant of the latter regiment had fallen and Berry seized the colors and bore them up the hill with his own. The illustrated press gave some attention to this exploit at the time, but no proper recognition of it has as yet been made. Sergeant Berry's character as a soldier had been formed long before this event, and his reputation for daring was already well established. He entered the service in 1867 and when he carried that flag up San Juan was filling out his thirty-first year in the service. All this time he had passed in the cavalry and had engaged in many conflicts with hostile Indians and ruffians on our frontiers.

Perhaps the most important parts taken by any enlisted men in the cavalry division were those taken by Sergeants Foster and Givens. The former was First Sergeant of Troop G and as the troop was making its way to the hill by some means the

Spaniards were able not only to discover them but also the direction in which they were moving and to determine their exact range. Sergeant Foster ventured to tell the Lieutenant in charge that the course of advance should be changed as they were marching directly into the enemy's guns.

"Silence," shouted the Lieutenant. "Come on, men; follow me." "All right, sir," said the Sergeant; "we'll go as far as you will." The next instant the Lieutenant was shot through the head, leaving Sergeant Foster in command. Immediately the troop was deployed out of the dangerous range and the Sergeant by the exercise of good judgment brought his men to the crest of the hill without losing one from his ranks. At the time of this action Sergeant Foster was a man who would readily command attention. Born in Texas and a soldier almost continuously since 1875, part of which time had been passed in an infantry regiment, he had acquired valuable experience. In 1888, while serving in the cavalry, he had been complimented in General Orders for skill in trailing raiding parties in Arizona. He was a resolute and stalwart soldier, an excellent horseman and possessed of superior judgment, and with a reputation for valor which none who knew him would question. The return of Troop G, Tenth Cavalry, for July, 1898, contains the following note: "Lieutenant Roberts was wounded early in the engagement; Lieutenant Smith was killed about 10.30 a. m. while gallantly leading the troop in the advance line. After Lieutenant Smith fell the command of the troop devolved upon First Sergeant Saint Foster, who displayed remarkable intelligence and ability in handling the troop during the remainder of the day. Sergeant Foster's conduct was such as cannot be excelled for valor during the operations around Santiago. He commanded the troop up the hills of San Juan."

Sergeant William H. Givens, of Troop D, Tenth Cavalry, also commanded in the action against San Juan. His Captain, who was wounded three times in the fight, being finally disabled before reaching the hill, makes the following report: "Sergeant William H. Givens was with the platoon which I commanded; whenever I observed him he was at his post exercising a steadying or encouraging influence on the men, and conducting himself like the thorough soldier that I have long known him to be. I understand to my great satisfaction that he has been rewarded by an appointment to a lieutenancy in an immune regiment."

The Descriptive list of Sergeant Givens, made on August 4th, 1898, contains these remarks:

"Commanded his troop with excellent judgment after his captain fell at the battle of San Juan, July 1, 1898, leading it up the hill to the attack of the blockhouse.

"Character: A most excellent soldier."*

Sergeant Givens may also be called an "old-timer." He had enlisted in '69, and had passed all that time in hard frontier service. The troop in which he enlisted during the years

---

*The major commanding the squadron in which Sergeant Givens' troops served, writes to the sergeant the following letter:

Sergeant William H. Givens, Troop D, 10th Cavalry, Fort Clark, Texas.

Sergeant:—When making my report as commander of the Second Squadron, 10th U. S. Cavalry, for action of July 1, 1898, at San Juan Hills, I did not mention any enlisted men by name, as I was absent from the regiment at the time of making the report and without access to records, so that I could not positively identify and name certain men who were conspicuous during the fight; but I recollect finding a detachment of Troop D under your command on the firing line during the afternoon of July 1st. Your service and that of your men at that time was most creditable, and you deserve special credit for having brought your detachment promptly to the firing line when left without a commissioned officer.

THEO. J. WINT,
Lieutenant-Colonel, 6th U. S. Cavalry.

Second Lieutenant, 10th Cavalry.

True copy:

1876-78 was almost constantly engaged with hostile Indians along the Mexican border, and Sergeant Givens was called upon to take part in numerous scouts in which there were many striking adventures. He was also in that memorable campaign against Victoria, conducted by General Grierson. Sergeant Givens was an ideal soldier and worthy the commendations bestowed upon him by his troop commander and others. Captain Bigelow received his disabling wound about seventy-five yards from the blockhouse and was taken to the rear under heavy fire by two soldiers of the troop by the name of Henderson and Boardman.

Lieutenant Kennington, reporting the work of the troop on that morning says that Corporal J. Walker was probably the first soldier to reach the top of the hill and is believed to have shot the Spaniard who killed Lieutenant Ord. The report containing the above statement is dated July 5, 1898. Since that time the matter has been fully investigated by Captain Bigelow and the fact ascertained that Corporal Walker did arrive first on the hill and did shoot the Spaniard referred to and he has been recommended for a Medal of Honor in consequence.

The Sergeant-Major of the Tenth Cavalry, Mr. E. L. Baker, who served with great credit during the Santiago campaign, is a soldier with an excellent record. He was born of French and American parentage in Wyoming and enlisted in the Ninth Cavalry as trumpeter in 1882, serving five years in that regiment. He then enlisted in the Tenth Cavalry, and in 1892 became Sergeant-Major. Being desirous of perfecting himself in the cavalry service he applied for an extended furlough with permission to leave the country, intending to enter a cavalry school in France. In this desire he was heartily endorsed by the officers of his regiment, and was specially commended by General Miles, who knew him as a soldier and who highly appreciated him as such. The breaking out of the Spanish war soon after he had made application prevented a full consideration of his case. In 1897 Sergeant-Major Baker published a specially valuable "Roster of the Non-Commissioned Officers of the Tenth U. S. Cavalry, with Some Regimental Reminiscences, etc.," which has been of marked service in the preparation of the sketches of the enlisted men of his regiment. He contributes the interesting sketch of his experiences in Cuba with his regiment, which follows this chapter, and which will prove to many perhaps the most interesting portion of my book.

The Twenty-fourth Infantry advanced in that line of attack on the extreme left and reached the crest of the San Juan Hills in such numbers as to lead the press correspondents and others to conclude that there were more men of this regiment promptly on the ground than of any other one regiment. It is certain they made a record for heroism in that assault as bright as any won on the field that day; and this record they raised to a magnificent climax by their subsequent work in the fever hospital at Siboney. For their distinguished service both in the field and in the hospital, the colored ladies of New York honored themselves in presenting the regiment the beautiful stand of colors already mentioned. As these fever-worn veterans arrived at Montauk they presented a spectacle well fitted to move strong men to tears. In solemn silence they marched from on board the transport Nueces, which had brought them from Cuba, and noiselessly they dragged their weary forms over the sandy roads and up the hill to the distant "detention camp." Twenty-eight of their number were reported sick, but the whole regiment was in ill-health.

These were the men who had risked their lives and wrecked their health in service for others. Forty days they had stood face to face with death. In their soiled, worn and faded clothing, with arms uncleaned, emaciated, and with scarce strength enough to make the march before them, as they moved on that hot 2nd of September from the transport to the camp, they appeared more like a funeral procession than heroes returning from the war; and to the credit of our common humanity it may be recorded that they were greeted, not with plaudits and cheers, but with expressions of real sympathy. Many handkerchiefs were brought into view, not to wave joyous welcome, but to wipe away the tears that came from overflowing hearts. At no time did human nature at Montauk appear to better advantage than in its silent, sympathetic reception of the Twenty-fourth Infantry.

Of these shattered heroes General Miles had but recently spoken in words well worthy his lofty position and noble manhood as "a regiment of colored troops, who, having shared equally in the heroism, as well as the sacrifices, is now voluntarily engaged in nursing yellow fever patients and burying the dead." These men came up to Montauk from great tribulations which should have washed their robes to a resplendent whiteness in the eyes of the whole people. Great Twenty-fourth, we thank thee for the glory thou hast given to American soldiery, and to the character of the American Negro!

Thus these four colored regiments took their place on the march, in camp, in assault and in siege with the flower of the American Army, the choice and pick of the American nation, and came off acknowledged as having shared equally in heroism and sacrifices with the other regular regiments so engaged, and deserving of special mention for the exhibition of regard for the welfare of their fellow man. The query is now pertinent as to the return which has been made to these brave men. The question of Ahasuerus when told of the valuable services of the Jew, Mordecai, is the question which the better nature of the whole American people should ask on hearing the general report of the valuable services of the Negro Regular in the Spanish War. When Ahasuerus asked: "What honor and dignity hath been done to Mordecai for this?" his servants that ministered unto him were compelled to answer: "There is nothing done for him." Looking over these four regiments at the time of this writing an answer somewhat similar in force must be returned. That the colored soldier is entitled to honor and dignity must be admitted by all who admire brave deeds, or regard the welfare of the state. The colored soldier, however, was compelled to stand by and see a hundred lieutenancies filled in the Regular Army, many in his own regiments, only to find himself overlooked and to be forced to feel that his services however valuable, could not outweigh the demerit of his complexion.

The sum total of permanent advantage secured to the colored regular as such, in that bloody ordeal where brave men gave up their lives for their country's honor, consists of a few certificates of merit entitling the holders to two dollars per month additional pay as long as they remain in the service. Nor is this all, or even the worst of the matter. Men who served in the war as First Sergeants, and who distinguished themselves in that capacity, have been allowed to go back to their old companies to serve in inferior positions. Notably is this the case with Sergeant William H. Givens, whose history has been detailed as commanding Troop D, Tenth Cavalry, after Captain Bigelow fell, and who heroically led the troop up the hill. He is now serving in his old troop as Corporal, his distinction having actually worked his reduction rather than substantial promotion.

It must not be inferred from the foregoing, however, that nothing whatever was done in recognition of the gallantry of the colored regulars. Something was done. Cases of individual heroism were so marked, and so numerous, that they could not be ignored. The men who had so distinguished themselves could not be disposed of by special mention and compliments in orders. Something more substantial was required. Fortunately for such purpose four regiments of colored United States Volunteer Infantry were then in course of organization, in which the policy had been established that colored men should be accepted as officers below the grade of captain. Into these regiments the colored men who had won distinction at Santiago were placed, many as Second Lieutenants, although some were given First Lieutenancies. This action of the Government was hailed with great delight on the part of the colored Americans generally, and the honors were accepted very gratefully by the soldiers who had won them on the field. Fortunately as this opening seemed, it turned out very disappointing. It soon became evident that these regiments would be mustered out of the service, as they had proven themselves no more immune, so far as it could be determined from the facts, than other troops. The Lieutenants who had been most fortunate in getting their commissions early got about six or seven months' service, and then the dream of their glory departed and they fell back to the ranks to stand "attention" to any white man who could muster political influence sufficient to secure a commission. Their day was short, and when they were discharged from the volunteer service, there appeared no future for them as commissioned officers. Their occupation was indeed gone. It was for them a most disappointing and exasperating promotion, resulting in some cases in loss of standing and in financial injury. Their honors were too short-lived, and too circumscribed, to be much more than a lively tantalization, to be remembered with disgust by those who had worn them. Cruel, indeed, was the prejudice that could dictate such a policy to the brave black men of San Juan. The black heroes, however, were not without sympathy in their misfortune. The good people of the country had still a warm place in their hearts for the colored soldier, despite the sayings of his maligners.

The people of Washington, D. C., had an opportunity to testify their appreciation of the Tenth Cavalry as that regiment passed through their city on its way to its station in Alabama, and later a portion of it was called to Philadelphia to take part in the Peace Jubilee, and no troops received more generous attention. To express in some lasting form their regard for the regiment and its officers, some patriotic citizens of Philadelphia presented a handsome saber to Captain Charles G. Ayres, who had charge of the detachment which took part in the Peace Jubilee, "as a token of their appreciation of the splendid conduct of the regiment in the campaign of Santiago, and of its superb soldierly appearance and good conduct during its attendance at the Jubilee Parade in Philadelphia."

Likewise when the Twenty-fifth Infantry arrived at its station at Fort Logan, Colorado, the people of Denver gave to both officers and men a most cordial reception, and invited them at once to take part in their fall carnival. All over the country there was at that time an unusual degree of good feeling toward the colored soldier who had fought so well, and no one seemed to begrudge him the rest which came to him or the honors bestowed upon him.

This state of feeling did not last. Before the year closed assiduous efforts were made to poison the public mind toward the black soldier, and history can but record that these efforts were too successful. The three hundred colored officers became an object at which both prejudice and jealousy could strike; but to reach them the reputation of the entire colored contingent must be assailed. This was done with such vehemence and persistency that by the opening of 1899 the good name of the black regular was hidden under the rubbish of reports of misconduct. So much had been said and done, even in Denver, which had poured out its welcome words to the heroes of El Caney, that the Ministerial Alliance of that city, on February 6, 1899, found it necessary to take up the subject, and that body expressed itself in the unanimous adoption of the following resolutions:

RESOLUTIONS ADOPTED UNANIMOUSLY BY THE MINISTERIAL ALLIANCE OF DENVER, FEBRUARY 6, 1899.

Resolved, By the Ministerial Alliance of the City of Denver, that the attempt made in certain quarters to have the Twenty-fifth Regiment, United States Infantry, removed from Fort Logan, appears to this body to rest on no just grounds, to be animated on the contrary by motives unworthy and discreditable to Denver and the State, and that especially in view of the heroic record of the Twenty-fifth Regiment, its presence here is an honor to Denver and Colorado, which this Alliance would regret to have withdrawn.*

*Extract from *The Statesman,* Denver, after the departure of the 25th Infantry, and the arrival of the 34th:

Two policemen killed, the murderer at large and his comrades of the 34th Regiment busy boasting of their sympathy for him, and extolling his

The mustering out of the volunteers about the time this opposition was approaching what appeared to be a climax, causing the removal from the service of the colored officers, appeased the wrath of the demon, and the waves of the storm gradually sank to a peace, gratifying, indeed, to those who shuddered to see a black man with shoulder-straps. As the last Negro officer descended from the platform and honorably laid aside his sword to take his place as a citizen of the Republic, or a private in her armies, that class of our citizenship breathed a sigh of relief. What mattered it to them whether justice were done; whether the army were weakened; whether individuals were wronged; they were relieved from seeing Negroes in officers' uniforms, and that to them is a most gracious portion. The discharge of the volunteers was to them the triumph of their prejudices, and in it they took great comfort, although as a matter of fact it was a plain national movement coming about as a logical sequence, entirely independent of their whims or wishes. The injustice to the Negro officer does not lie in his being mustered out of the volunteer service, but in the failure to provide for a recognition of his valor in the nation's permanent military establishment.

---

deed to the skies, yet not a single petition has been prepared to have the regiment removed. The 25th Infantry, with its honor undimmed by any such wanton crime, with a record unexcelled by any regiment in the service, was the target for all sorts of criticism and persecution as soon as it arrived. The one is a white regiment, composed of the scum of the earth, the other a black regiment composed of men who have yet to do one thing of which they should be ashamed. Yet Denver welcomes the one with open arms and salutes with marked favor, while she barely suffered the other to remain.

Had it been a negro soldier who committed the dastardly deed of Saturday night the War Department would have been deluged with complaints and requests for removal, but not a word has been said against the 34th. Prejudice and hatred blacker than the wings of night has so envenomed the breasts of the people that fairness is out of the question. Be he black, no matter how noble and good, a man must be despised. Be he white, he may commit the foulest of crimes and yet have his crimes condoned.

The departure of the colored man from the volunteer service was the consequent disappearance of the colored military officer, with the single exception of Lieutenant Charles Young of the Regular Cavalry, had a very depressing effect upon the colored people at large, and called forth from their press and their associations most earnest protests. With a few exceptions, these protests were encouched in respectful language toward the President and his advisers, but the grounds upon which they were based were so fair and just, that right-thinking men could not avoid their force. The following resolution, passed by the National Afro-American Council, may be taken as representative of the best form of such remonstrance:

"*Resolved,* That we are heartily grieved that the President of the United States and those in authority have not from time to time used their high station to voice the best conscience of the nation in regard to mob violence and fair treatment of justly deserving men. It is not right that American citizens should be despoiled of life and liberty while the nation looks silently on; or that soldiers who, with conspicuous bravery, offer their lives for the country, should have their promotion result in practical dismissal from the army."

The nation graciously heeded the call of justice and in the re-organization of the volunteer army provided for two colored regiments, of which all the company officers should be colored men. Under this arrangement many of the black heroes of Santiago were recalled from the ranks and again restored to the positions they had won. Thus did the nation in part remedy the evil which came in consequence of the discharge of the volunteers, and prove its willingness to do right. Triumphantly did the Administration vindicate itself in the eyes of good people, and again did it place its withering disapproval upon the conduct of those who were ready to shout their applause over the worthy black officer's accidental humiliation. The Negro officer disappeared from the United States' Regiments as a Lieutenant only; but he returns to the same, or rather, to a higher grade of the same form of regiments, both as Lieutenant and Captain. How rapid and pronounced has been the evolution! It is true the Negro officer is still a volunteer, but his standing is measurably improved, both because of the fact of his recall, and also because the regiments which he is now entering have some prospect of being incorporated into the Regular Army. It does not seem probable that the nation can much longer postpone the increase of the standing army, and in this increase it is to be hoped the American Negro, both as soldier and officer, will receive that full measure of justice of which the formation of the present two colored regiments is so conspicuous a part.

DIARY OF E. L. BAKER, SERGEANT-MAJOR TENTH U. S. CAVALRY.

Appointed First Lieutenant Ninth U. S. Volunteer Infantry, and later Captain of the Forty-ninth Volunteer Infantry—Now Lieutenant in Philippine Scouts.

A TRIP FROM MONTANA TO CUBA WITH THE TENTH U. S. CAVALRY.

April 16, 1898, at 10.45 p. m., telegram was received from Department Headquarters, St. Paul, Minnesota, ordering the regiment to the Department of the Gulf.

As every click of the telegraph instrument was expected to announce a rupture in the diplomatic relations between the United States and the Kingdom of Spain, all knew that the mobilization of the army South meant preparing it for the serious work for which it is maintained.

On April 19 we were off for Chickamauga Park. En route we were heartily greeted. Patriotism was at its height. Every little hamlet, even, had its offerings. To compare the journey with Cæsar's march of triumph would be putting it mildly.

We arrived at the historic point April 25. Every moment of our stay there was assiduously devoted to organizing, refitting and otherwise preparing for the inevitable. Officers were sent to many parts of the country .to secure recruits Many also gave up details nd relinquished their leaves of absence to take part in the impending crisis.

May 14. We were moved a little nearer the probable theatre of operations. On account of some deficiency in water for troops at Tampa, the regiment was stopped at Lakeland, 30 miles this side, where many recruits were received; Troops increased to war strength, and new Troops established. Drills and instructions were also constantly followed up.

June 6. Orders were received to prepare headquarters, band and eight Troops dismounted, with trained men only, for service in Cuba. Recruits to be left in camp with horses and property.

June 7. We were off for Port Tampa, where the regiment embarked on the steamship Leona that afternoon.

June 8. She steamed from the dock. When the expedition seemed to be forming, news was received that the dreaded Spanish fleet was being sighted, evidently lying in wait for army transports. So we steamed back to the pier. Many of the men appeared disappointed at the move, probably not realizing that there was too much water in the Atlantic Ocean for the 5th Army Corps to drink.

To my mind, the Divine Providence surely directed the move, as the delay enabled the force to be swelled several thousand, every one of whom was needed before Santiago.

June 14. We steamed out of Tampa Bay, amid cheers and music from the thirty odd transports, heavily escorted by naval vessels. Among them were the much talked-of dynamiter, Vesuvius, and the beautiful little cruiser, Helena. Off Dry Tortugas that formidable warship, Indiana, joined the fleet.

Splendid weather; nothing unusual transpiring, though our transport, which also contained the First U. S. Cavalry, had a seemingly close call from being sent to the bottom of the sea, or else being taken in as a prisoner, which the enemy could have done with impunity.

Whilst going down the Saint Nicholas Chanel, in Cuban waters, the vessel was deliberately stopped about midnight, June 16, and left to roll in the trough of the sea until the morning of the 17th, in consequence of which we were put 20 hours behind the fleet and without escort, almost in sight of the Cuban shores.

Men were indignant at having been placed in such a helpless position, and would have thrown the captain of the ship, whom they accused of being a Spanish sympathizer and otherwise disloyal, overboard without ceremony, but for the strong arm of military discipline. We were picked up by the U. S. Cruiser Bancroft, late in the afternoon, she having been sent in quest of the Jonah of the fleet. Upon approach of the ship there were prolonged cheers from all of Uncle Sam's defenders. The only explanation that I have ever heard for this unpardonable blunder on the part of the ship's crew was that they mistook a signal of a leading vessel.

June 20. Land was sighted.

June 21. Dispatch boats active; transports circling; Morro Castle pointed out; three days' rations issued to each man; no extra impedimenta to be taken ashore; crew preparing for landing.

June 22. As we neared Daiquiri, the designated place for disembarking, flames could be seen reaching almost to the heavens, the town having been fired by the fleeing Spaniards upon the approach of war vessels of Sampson's fleet, who were assembling to bombard the shore and cover our landing. After a fierce fire from these ships, the landing was effected with loss of two men of our regiment, who were doubtless crushed to death between the lighters. They were buried near the place of recovery the next morning.

The few half-clothed and hungry-looking natives on shore seemed pleased to see us. Daiquiri, a shipping point of the Spanish-American Iron Company, was mostly deserted. The board houses seemed to have been spared, while the sun-burned huts thatched with palm were still smoking, also the roundhouse in which there were two railroad locomotives, warped and twisted from the heat. The Spanish evidently fired everything they could before evacuating.

June 23. At 6.00 p. m. Troops A, B, E and I, left with four Troops of the First U. S. Cavalry and Rough Riders (First U. S. Volunteer Cavalry) as advance guard of the Army of Invasion on the main road to Santiago de Cuba; about 800 men all told, three Hotchkiss guns, manned by ten cavalrymen, accompanied also by the Brigadier Commander, General S. M. B. Young and staff.

NOTE.—These troops marched about 13 miles through a drenching rain from 7 to 10 p. m.; bivouacked one hour later. On the 24th, after breakfast, took the trail about 5.15 a .m. The vapor from wet clothing rose with the sun, so that you could scarcely recognize a man ten feet away. About three and one-half miles above Siboney the command was halted; the first U. S. Volunteer Cavalry (Rough Riders) sent to the left; proceeding farther about one mile, the main column was split, First U. S. Cavalry going to the right, the Tenth Cavalry remaining in the center. General Wheeler joined at this point, accompanied by his orderly, Private Queene, Troop A, Tenth Cavalry. Disposition of the troops was explained by General Young, who had located his headquarters with the Tenth U. S. Cavalry; General Wheeler made his the same. Hotchkiss guns were ordered closed up; magazines filled. The column had proceeded but a short way when the engagement opened in all its fury; troops were deployed and advanced in the direction from which the bullets were coming the thickest, as rapidly as the formation of the ground would permit, the left of the line touching the right of the Rough Riders.

June 24. Headquarters, band and the remainder of the First and Tenth U. S. Cavalry were off at 6 A. M. The road was alive with troops (C, D, F, G,) colonels and privates alike lugging their rations and bedding beneath that ever watchful tropical sun, feeling as though they would wilt at every step, the undergrowth being so thick and tall that scarcely any breeze could get to you.

On emerging from this thicket, through which we had been marching for several hours, the Sampson fleet could be heard firing on the Spanish batteries on shore. Marines and other troops could be seen crossing the mountains above Altares; this revived the men very much. As we approached Verni Jarabo (Altares?), we were met by General Lawton, who informed our Colonel that the advance guard was engaged with the Spanish at La Guasima, and that it was hard pressed. Our pace was quickened; the news appeared to lighten our heavy packs as we toiled to the front to assist our comrades. The roar of the artillery became plainer; wounded men along the road as well as those played-out from the intense heat. Women and children were fleeing to places of safety. Our forces were repairing a railroad engine and track; also tearing up a piece leading to a Spanish blockhouse. In fact, everything seemed to have on an exceedingly warlike tint, but our advance continued as swifty as our weary feet would allow, which soon brought us to a number of our own comrades conveyed on litters from La Guasima, where our advance guard was tussling hard with the Dons for the honors of the day.

Upon arrival of reinforcements, victory had been wrested from the Dons fairly by the advance guard without assistance. Every one greeted each other, as though it had been a year instead of a few hours since parting. The First U. S. Cavalry and Rough Riders were unstinted in extolling the fighting qualities of their brothers in arms, the Tenth U. S. Cavalry.

The enemy was struck early June 24, entrenched on the heights of La Guasima, near Sevilla, on the main road from Daiquiri to the city of Santiago de Cuba. The advance guard was soon hotly engaged with them; after a very desperate fight of over one hour, the enemy was driven in confusion from their intrenchments. Our men were too exhausted to follow them. The Tenth Cavalry lost 13 killed and wounded. For a while it was a terrific fight, as the enemy was strongly intrenched on the heights and our men had to climb them subjected to their fire, which was very accurate, and much of it doubtless from machine guns in hands of experienced men. Our men had also to contend with the thickest underbrush, wire fences (the famous military trochas) and Spanish daggers jabbing them in side at every step. For a while the situation was serious. The decisive blow of the attack seems to have been struck at an opportune moment, and the enemy withdrew in confusion.

It has been estimated that about 4,000 Spanish were engaged. Everything indicated that they lost heavily; a Santiago paper put it at 240. The writer and the Sergeant-Major of the First U. S. Cavalry superintended the digging of one large grave where all the dead of the two regiments were interred according to the Episcopal service. The Rough Riders, being farther to our left, buried their own. If advantage of position goes for anything, the Spanish should have annihilated the Americans as they approached the stronghold.

The command remained on the battlefield until June 26, when it proceeded to Sevilla, an old coffee and sugar plantation, to await the assembling of the army and placing of the artillery.

Our camp at Sevilla was an interesting one in many ways. It was pitched between the main road and a stream of excellent water. From the hill beyond, the Spanish works could be viewed. From the roadside many acquaintances were seen, also generals, foreign military attaches, troops, artillery and pack trains. Wheeled transportation seemed entirely out of its place in Cuba; one piece of artillery was noticed with 24 horses tugging away at it.

The Cuban Army, cavalry and infantry, passed us at this point, which seemed to consist of every male capable of swelling the crowd. Those unable to carry or secure guns had an old knife or machete strapped to them.

On June 30, about 4 P. M., shortly after our daily shower, which was a little more severe and much longer than usual, the regiment was put in motion for the front. We had marched about 1600 yards when the war balloon was seen ascending some distance to our right. As the balloon question was new, every one almost was stumbling on the man's heels in front, trying to get a peep at this wonderful war machine.

After much vexatious delay, narrow road crowded with troops, a pack train came along and added its mite to the congestion, as some of the mules turned their heels on the advancing column when pushed too much.

We finally merged into a beautiful lawn, site of the Division Hospital, where all were as busy as beavers in placing this indispensable adjunct in order. Here the work of July 1 was clearly suggested. Proceeding, wading and rewading streams, we bivouacked beyond the artillery on the heights of El Poso, an old sugar plantation, about four miles off, in plain view of the city of Santiago. The lights of the city showed so brightly, the enemy offering no resistance to our advance, I could not help feeling apprehensive of being in a trap. I thought so seriously over the matter that I did not unroll my pack, so as to be ready at an instant. Simply released my slicker, put it on, and lay down where I halted.

Early July 1 all the brigade was up, getting breakfast and making as much noise as if on a practice march. The Tenth Cavalry did not make any fire until orders were received to that effect. I remarked to my bunky that we were not going to fight evidently, as the smoke would surely disclose our presence and enable the enemy's artillery to get our range The whole of Santiago seemed to be decorated with hospital flags.

At 6.30 a shell from Capron's battery, U. S. Artillery, directed at a blockhouse in El Caney, announced that the battle was on. Then the musketry became general. All stood and watched the doomed village quite a while as the battle progressed. Soon Grices' battery of the U. S. Artillery, which was in support, belched forth destruction at the Spanish works of the city, using black powder. The fire was almost immediately returned by the enemy's batteries, who had smokeless. They were shortly located when a fierce duel took place. The Dons were silenced, but not until we had suffered loss. During this fire an aide—Lieut. Wm. E. Shipp, Tenth Cavalry, Brigade Quartermaster—brought orders for us to take position on the left of the First U. S. Cavalry. The line extended nearly north and south on a ridge some three or four miles from the city, where the regiment was exposed to much of the return fire from the enemy's batteries. The men exhibited no special concern and watched the flight of the death messenger as eagerly as if at a horse race. Adjutant Barnum here divided the band and turned it over to the surgeons to assist in caring for the wounded, and directed Saddler Sergeant Smith and myself to accompany the Colonel in advance. When Lieut. Shipp delivered his orders, some of the officers remarked, "You are having a good time riding around here." He replied that it was no picnic riding among bullets, and that he would prefer being with his troops.

After the artillery had ceased firing, the regiment moved to the right, passed El Poso, where there were additional signs of the enemy's havoc among our troops, proceeded down the road leading to Santiago. The movement of the regiment was delayed as it approached the San Juan River, by an infantry brigade which had halted.

The regiment came within range of musket fire about three-quarters or one-half mile from the crossing. Upon reaching the ford the Colonel (Baldwin) rode nearly across the stream (closely followed by his regiment) when we were greeted by the Dons with a terrific volley of musketry, soon followed by artillery, which caused us to realize more fully than ever, that "things were coming our way." Orders were given to throw off packs and get cover. In removing his, Sergeant Smith, on my immediate left, was assisted by a Spanish bullet, and an infantry soldier fell as my pack was thrown off to the right. In seeking cover men simply dropped to the right and left of the road in a prone position.

The regiment was here subjected to a terrific converging fire from the blockhouse and intrenchments in front and the works further to the left and nearer the city. The atmosphere seemed perfectly alive with flying missiles from bursting shells over head, and rifle bullets which seemed to have an explosive effect. Much fire was probably drawn by the war balloon, which preceded the regiment to a point on the edge of the river, near the ford, where it was held. This balloon undoubtedly rendered excellent service in locating positions of the Spanish works and developing an ambush which had been laid for us, but the poor, ill-fated balloon certainly received many uncomplimentary remarks during our stay in its vicinity.

It seemed as though the Spanish regarded the balloon as an evil agent of some kind, and as though every gun, both great and small, was playing on it. I made several trips under it following the Colonel, who repeatedly rode up and down the stream, and I would have been fully satisfied to have allowed my mind even to wander back to the gaily lighted ball rooms and festivals left behind only a few months before.

While on the last trip under the balloon a large naval shell exploded, knocking the Colonel's hat off, crippling his horse, and injuring the rider slightly in the arm and side, all of course, in addition to a good sand bath. I then joined the regiment, some rods beyond, then under cover. In crouching down behind a clump of brush, heard some one groan; on looking around, saw Private Marshall struggling in the river wounded. Immediately rushing to his assistance another of those troublesome shells passed so close as to cause me to feel the heat. It did not stop the effort, however, and the wounded man was placed in safety.

The regiment remained in the road only a few moments when it was ordered to take position behind the river bank some yards above the balloon for protection; while moving to that position, and while there, suffered much loss. Why we did not lose heavier may be attributed to the fact that the enemy's musket fire was a trifle high, and their shells timed from one-half to one second too long, caused them to explode beyond, instead of in front, where the shells would have certainly secured the Dons' maximum results, as, after the balloon was cut down, you could scarcely hold your hand up without getting it hit. During the battle, one trooper fell upon a good-sized snake and crushed it to death, and another trooper allowed one of these poisonous reptiles to crawl over him while dodging a volley from the Spanish Mausers.

The shrapnel and canister shells, with their exceedingly mournful and groaning sound, seemed to have a more terrifying effect than the swift Mauser bullet, which always rendered the same salutation, "Bi-Yi." The midern shrapnel shell is better known as the man-killing projectile, and may be regarded as the most dangerous of all projectiles designed for taking human life. It is a shell filled with 200 or 300 bullets, and having a bursting charge, which is ignited by a time fuse, only sufficient to break the base and release the bullets, which then move forward with the velocity it had the time of brusting. Each piece is capable of dealing death to any living thing in its path. In practice firing, it is known where, by one shot, 152 hits were made by a single shrapnel. In another,

215 hits are recorded. Imagine then, the havoc of a well-directed shrapnel upon a group of men such as is here represented. Capron's battery at El Caney cut down 16 cavalrymen with one shell.

After a delay of about 30 minutes, during part of the time, the writer, assisted by Sergeant Smith and Mr. T. A. Baldwin, cut all the wire fences possible. Mr. Baldwin was dangerously wounded while so engaged just before the general advance.

The regiment merged into open space in plain view of and under the fire of the enemy; and formed line of battle facing toward the blockhouses and strong intrenchments to the north, occupied by the Spanish, and advanced rapidly in this formation, under a galling, converging fire from the enemy's artillery and infantry, on the blockhouses and heavy intrenchments to the right front. Many losses occurred before reaching the top of the hill, Lieut. W. H. Smith being killed while gallantly conducting his troop as it arrived on the crest. Lieut. W. E. Shipp was killed about the same instant, shortly after leaving Lieutenant Smith, further to the left and near the pond on the sunken road leading to Santiago. Lieutenant Smith was struck in the head and perished with a single groan. Lieutenant Shipp was hit near the heart; death must have been almost instantaneous, though it appears he made an effort to make use of his first aid package. Thus the careers of two gallant and efficient officers whose lives had been so closely associated were ended.

Private Slaughter, who was left in charge of Lieutenant Smith's body, was picked off by the Spanish sharpshooters, and Private Jackson, Lieutenant Shipp's orderly, was left as deaf as a post from a bursting shell.

The enemy having been driven back, northwest, to the second

and third blockhouses, new lines were formed and a rapid advance made upon them to the new positions. The regiment assisted in capturing these works from the enemy, and planted two sets of colors on them, then took up a position to the north of the second blockhouse. With some changes in position of troops, this line, one of the most advanced, about three hundred yards of the enemy, was held and intrenchments dug under a very heavy and continuous fire from the Spanish intrenchments in front, July 2 and 3.

In their retreat from the ridge, the enemy stood not on the order of their going, but fled in disorder like so many sheep from the scene, abandoning a quantity of ammunition, which was fired at them subsequently from our rapid-fire guns. Our men were too exhausted to pursue them, footwear and clothing being soaked by wading rivers, they had become drenched with rain, and when they reached the crest they were about played-out; having fought about 12 hours, most of which was under that ever-relentless tropical sun.

Throughout the night, work on the intrenchments was pushed, details buried the dead, improvised litters, and conveyed the wounded to hospitals, all of which was prosecuted with that vim for which the regular soldier is characterized, notwithstanding their water-logged condition.

The regiment acted with extraordinary coolness and bravery. It held its position at the ford and moved forward unflinchingly after deployment, through the dense underbrush, crossed and recrossed by barbed wire, under heavy and almost plunging fire from the Spanish works, while attacking with small arms an enemy strongly posted in intrenchments and blockhouses, supported by artillery, and who stubbornly contested every inch of ground gained by the American troops.

Officers were exceedingly active and tireless in their efforts to inspire and encourage the men. You could hear them call out, "Move right along; the Spaniards can't shoot; they are using blanks." One officer deliberately stopped and lit his pipe amid a shower of bullets, and then moved on as unconcerned as if on target practice.

The rifle pits occupied by the enemy were intrenchments in reality, dug almost shoulder deep, and faced with stone, being constructed without approaches, leaving the only avenue for escape over the parapet, which was equivalent to committing suicide, in face of the unerring marksmanship of the United States troops.

We were afterward told by a Spanish soldier how they were held in these trenches by an officer stationed at each end with a club; also how they depended on their officers for everything. This may account for the large percentage of our officers picked off by the Dons. I observed during the battle that when spotted by the enemy, delivering orders or busying about such duties as usually indicated some one in authority, the Spanish would fire whole volleys at an individual, this evidently with a view to demoralizing the rank and file by knocking off the officers.

The Spanish also tried an old Indian trick to draw our fire, or induce the men to expose themselves, by raising their hats on sticks or rifles, or placing them upon parapets, so when we went to fire they would aim to catch us as we rose with a terrific volley. The Dons were, however, soon convinced of their folly in this respect, as we always had a volley for the hats and a much stouter one for the enemy as he raised to reply to the volley at the hats. The Tenth Cavalry had fought Indians too long in the West to be foiled in that manner.

We were annoyed much by the Spanish sharpshooters stationed in tops of the beautiful palms and other trees of dense foliage. A number of these guerillas were found provided with seats, water and other necessaries, and I am told some of them had evidently robbed our dead to secure themselves an American uniform, that they might still carry on their nefarious work undetected.

Many of the disabled received their second and some their mortal wound, while being conveyed from the field by litter-bearers.

Though it was the tendency for a time to give the sharp-shooter story little or no credence, but to lay the matter to "spent bullets"; it seemed almost out of the question that "spent bullets" should annoy our Division Hospital, some four or five miles from the Spanish works. It would also seem equally as absurd that a bullet could be trained to turn angles, as several of our men were hit while assembled for transfer to general hospital and receiving temporary treatment at the dressing station located in an elbow of the San Juan River.

The Division Hospital was so harassed that it was necesasry to order four Troops of the 9th U. S. Cavalry there for guard. While en route to the hospital on the morning of July 2 with wounded, I saw a squad of the 2nd U. S. Cavalry after one of these annoying angels, not 20 feet from the road. On arrival at the hospital I was told by a comrade that several had been knocked from their stage of action. On July 1, our Color-Sergeant was shot from a tree after our line had passed beneath the tree where he was located. July 3, three more fell in response to a volley through tree tops, and on July 14, while waiting the hand to reach the hour for the bombardment of the city, one of the scoundrels deliberately ascended a tree in plain view of, and within two hundred yards of, our line. It was a good thing that the white flag for surrender appeared before the hour to commence firing, otherwise Spain would have had at least one less to haggle with on account of back pay.

To locate a sharpshooter using smokeless powder among the dense tropical growth may be compared with "looking for a needle in a haystack."

The killed and wounded in battle present a scene well calculated to move the most callous. Men shot and lacerated in every conceivable manner; some are expressionless; some just as they appeared in life; while others are pinched and drawn and otherwise distorted, portraying agony in her most distressful state. Of the wounded, in their anguish, some are perfectly quiet; others are heard praying; some are calling for their mothers, while others are giving out patriotic utterances, urging their comrades on to victory, or bidding them farewell as they pass on to the front. July 1, in passing a wounded comrade, he told me that he could whip the cowardly Spaniard who shot him, in a fair fist fight.

During the first day's battle many interesting sights were witnessed. The new calibre 30 Gatling guns were in action. These cruel machines were peppering away several hundred shots each per minute and sweeping their front from right to left, cutting down shrubbery and Spaniards like grain before the reaper. I observed the excellent service of the Hotchkiss Mountain gun; they certainly do their work to perfection and well did the Dons know it. Many shots fired into the "blind ditches and blockhouses" of the enemy caused them to scatter like rats. These guns use a percussion shell nearly two inches, and can be packed on mules. They were designed for light service with cavalry on the frontier. Four of these little beauties were manned by men of the Tenth Cavalry. The Spanish made it so hot for the boys that they would have to roll the gun under cover to load, and then steal it back to fire.

I saw one of our light batteries of artillery go in position under fire at the foot of San Juan Hill. The movement was swiftly and skillfully executed. A most interesting feature of this was to see the Caissons, drawn by six magnificent horses, off for ammunition. Three drivers to each outfit, one to each pair of horses; all plying the whip at every jump, would remind you of a Roman chariot race coming around on their last heat.

Wheeled vehicles of war suffer more than other troops, on account of their stationary positions. It is here that the dreaded sharpshooter comes in for glory, by picking off the gunners and other individuals.

Pack trains were seen dashing along the line with that always absolutely essential—ammunition—thereby gladdening the hearts of the boys who were doing their utmost to expend every round in their belts to gain another foot of Spanish territory.

During all these stirring events the stomachs of the real heroes were not neglected, and most certainly not along our part of the line. Pack mules were brought right up to the line under a hot fire, loaded with sugar, coffee, bacon and hardtack, all of which was in plenty. Some of the mules were killed and wounded, but this did not retard the advance of the train. When near the firing line some one called, "Whose rations?" A prompt reply, "Hungry soldiers."

The daring horseman was all that was needed to make the situation complete. Without participation of cavalry, the ideal warrior disappears from the scene, and the battle and picture of war is robbed of its most attractive feature.

Late in the afternoon, July 1, I was directed to take Saddler Sergeant Smith and bring to the firing line all the men I could find of the regiment. Going to the dressing station, collected those who had brought or assisted wounded there, thence across a portion of the field passed over a few hours previous. Men were found almost exhausted, soaking wet, or a solid mass of mud, resting as comfortably as if in the finest of beds; many of them had been on picket duty all night before, to which was added the hard day's work not then completed. After locating all I could, we went to the crest of the San Juan Hill, to the left of the sunken road, where the First U. S. Cavalry was reforming, and there picked up a few more who had joined that regiment.

The Tenth Cavalry having in the meantime taken another position, I set out to find it, going in front, telling Smith to bring up the rear. We were detained a short time near Sunken Roads by shells from Cervera's fleet, which were falling in it at a lively rate. Barbed wire prevented us from "running the gauntlet." Shortly after crossing the road an officer passed us, his horse pushed to his utmost, telling us to take all the ammunition that we possibly could on the firing line. About that instant, the pack train came thundering by, which we relieved of a few thousand rounds in short order. I was much amused at one of the men who innocently asked, "Where are we to get axes to burst these strong boxes?" The job was speedily accomplished before the boxes were on the ground good, and most certainly in less time than it would have taken to explain matters to the inexperienced. We were soon off again, tramping all over the country, through darkness, running into wire entanglements, outposts and pickets, and within fifty yards of the enemy (subsequently ascertained).

About 11.00 P. M. found Colonel Roosevelt a few hundred yards from the Spanish lines with some of my regiment, the First Cavalry, and Rough Riders, at work on trenches, where we reported. All seemed glad to have my little reinforcement, about 65 men, and ammunition. I never felt so relieved at anything as I did to get that herculean task off my hands, a job as hard as working a problem in the third book of Euclid. The men were so tired that they would lie down at every stop to find the right road or the way out of the wire entanglements constantly encountered. I have never seen in a book anything to equal the Spanish wire entanglements. Barbed wire was stretched in every nook and corner, through streams, grass, and from two inches to six feet in height, and from a corkscrew to a cable in design. It takes the nerve of a circus man to get men along when they are so exhausted that every place feels alike to them, and that they would gladly give away Mr. Jim Hill's fortune if they possessed it, for a few hours' sleep.

On arrival at the front, lunch was about over or just ready. Lieutenant E. D. Anderson (10th Cavalry) gave me two and one-half hardtacks from his supply, which he carried in his bosom. I was soon down for a little rest; all desultory firing had ceased; the pick and the shovel were the only things to disturb the quietude of that anxious night. Had been down but a short time when aroused by one of the Rough Riders, who had some rice and meat in an ammunition box which he brought from the captured blockhouse. The meat was undoubtedly mule, as the longer I chewed it the larger and more spongy it got, and were it not for the fact that I had had some experience in the same line many years before in Mexico while in pursuit of hostile Indians, I would certainly have accused our best friends (Rough Riders) of feeding us rubber. I made another effort for a little sleep, and was again aroused by some one passing around hardtack, raw bacon, etc., with instructions as to where to go to cook it. I thanked him and carefully laid it aside to resume my nap. At 2.40 A. M. the pickets were having such a lively set to, that I thought the general engagement was on. It was at this time I discovered that I was shivering cold, and that my teeth were rattling equal to a telegraph sounder; so under the circumstances, I concluded not to try for any more sleep. The dew was falling thick and heavy; no coat, no blanket, top shirt torn in strips from the brush, and undershirt wet and in my pack, thrown off on coming into battle.

Early July 22nd the artillery took position on our left. Pickets kept up firing from 2.40 A. M. until 5.25, when the engagement became general. Shortly after 6.00 A. M. our artillery opened on the Spanish works, who promptly returned the compliment. During the firing the Dons exploded a shell in the muzzle of one of our pieces. Adjutant Barnum fell at 6.30 A. M.; his wound was promptly dressed, when I started to the Division Hospital with him. Though seriously hurt, I have never seen a better natured man. While en route, we laid him down to eat a can of salmon *found in the road.* In response to his query, "What's up, Sergeant?" the salmon was passed him; he helped himself, no further questions were asked, and the journey was resumed. On arrival at the hospital he was quickly examined and placed on a comfortable cot. Many of the attendtns were completely played-out from overwork.

A visit to a field hospital will have a lasting place in your memory. Every way you turn, amid the cries and groans, you get a beck or call to ease this, or hand me that, and one feels badly because of his inability to extend them material aid in their sufferings.

On returning to the front, I found the regiment as hotly engaged as when I left it some hours before. As the fighting was from trenches, many of our men were wounded by shells. Sharpshooters were on hand as usual. I was sent to the Captain of Troop E, under the crest of the hill, with orders to dig an approach to one of the enemy's trenches, evacuated the day before; also to bury some of their dead. While delivering the order, it being necessary to get very close on account of the noise, one of those ever vigilant sharpshooters put a bullet between our faces. The Captain asked me to cut the wire fence so his troops could get through more rapidly; while telling me, another bullet passed so close as to disturb the Captain's mustache. He took it good-naturedly, only remarking as he smiled, "Pretty close, Sergeant-Major!"

Firing ceased about 8 P. M. After all had had supper we changed position further to the right, where work on trenches was resumed. About 10.30 P. M. the Spaniards made an attack upon our lines, and I have never before or since seen such terrific firing; the whole American line, which almost encircled the city, was a solid flame of fire. The enemy's artillery replied, also their much-praised "Mausers," but to no avail; they had opened the ball, but Uncle Sam's boys did not feel like yielding one inch of the territory so dearly bought.

About midnight all hands were aroused by the dynamite cruiser Vesuvius "coughing" for the Dons. The roar was so great that it seemed to shake the whole island. To the uninitiated it would appear that some one had taken a few mountains several miles up in a balloon and thrown them down.

July 3. Firing by pickets commenced very early, and quite heavy, at 5.40 A. M. Terrific cannonading to the seaward was heard between 9 and 10 A. M. As there was some talk of the enemy making a sortie, all eyes were open. Dirt began falling in the pits from the jar, bells could be heard tolling in the city, and steam whistles in the harbor. There was much speculation as to what was in progress. I'll say that there were many glad hearts when the news reached us that *Sampson's fleet was King of the Seas.* At 12 M. all firing was ordered off, for flag of truce to enter the Spanish lines. When the order for cease firing was given, one of the troopers laid his gun upon the parapet and remarked that he "would not take $2000 for his experience, but did not want a cent's worth more." Work on bomb-proofs and breast works was continued incessantly until news of the surrender reached us.

July 4. Flag of truce all day; national and regimental colors placed on parapets. At noon the regiment paraded, and all hearts cheered by the patriotic telegram of the Commander-in-Chief—His Excellency, President McKinley. Refugees, in droves, could be seen leaving for several days, notice of bombardment having been served on the city.

July 5. There was much excitement when Lieutenant Hobson and party crossed our lines.

During truce, the monotony was broken occasionally by the presence of Spanish soldiers in quest of something to eat or desiring to surrender.

Truce was off July 10 at 4 P. M. Bombardment of the city commenced by the army and navy combined, which continued until 2 P. M. 11th. Gatling, dynamite, rapid-firing and Hotchkiss guns were so well trained that the Dons scarcely dared to raise their heads, and their firing was soon silenced. During the attack our part of the line suffered no loss. While occupying these works, it was discovered that the gun of the enemy that *annoyed us most was quite near a large building covered with Red Cross flags.*

During the truce all of our dead were located and buried. It was sad, indeed, to see the vultures swarming like flies, when we knew so well their prey.

Though prepared to, several times, no shots were exchanged after July 11, and all was quiet until date of capitulation. The hardest rain ever witnessed, accompanied by terrific thunder and lightning, was on the last day of the engagement. Trenches were flooded and everything appeared as a sea.

July 17, at 9 A. M., the regiment, with the remainder of the army, was assembled over the trenches to witness the formal surrender of General Toral, with the Spanish forces. Owing to the dense tropical growth, and its similarity in color to their clothing, little or nothing could be seen, beyond the straw hats of the Dons, as they marched through the jungles. At 12 M., we were again placed in the same position, to salute "Old Glory" as she ascended over the Governor's palace in the city, which was told by Capron's battery U. S. Artillery. At the first shot, every individual tested his lungs to their fullest capacity, bands of music playing national airs.

Spanish soldiers were soon over our lines, trading off swords, wine, cigarettes and trinkets for hard tack and bacon. This soon ended, as there were positive orders against our fraternizing. The Spaniards were a fine looking lot of young men; though generally small in stature, and were very neat and clean, considering. The officers were an intelligent and dignified looking set. The Dons were away ahead on ammunition, and away behind on eatables. A few musty, hard tack, thrown in our trenches, were devoured like so much fresh beef, by so many hungry wolves.

Campaigning in the tropics entails many hardships, though unavoidable and only to be expected, in war. War is horrible in any aspect in which it may be viewed. Even those features of it intended to be merciful, are full of harshness and rigor; and after all, fighting is the easiest part.

As the capitulation was complete, and Santiago was our's, we were ordered to change camp to a more healthful locality, with a view to allowing the men to recuperate. While en route many refugees were met returning to the city, men and women, with the scantiest clothing imaginable; large children even worse—in a nude state—all were making signs for something to eat.

In passing through El Caney, filth of all descriptions was piled up in the streets; stock was seen standing inside dwellings with occupants; young and old were emaciated—walking skeletons; children with stomachs bloated to thrice their natural size—due to the unsanitary condition of the huts, so I was informed.

The bare facts are, that "half has never been told" regarding the true condition of the Cubans, and it is truly a God-send that "Uncle Sam" was not delayed another day in letting the Don's breathe a little of nature's sweetest fragrance of the nineteenth century—Civilization.

The portion of the island I saw appears to be a beautiful park deserted and laid waste by the lavish application of the torch for many years. Magnificent mansions, or dwellings, in ruins; habitation scant, except near towns.

There were no domestic animals, except a few for saddle purposes, nor were there crops to be seen. No use whatever appears to be made of the luxuriant pasturage and rich fields. Sugar houses and sheds on plantations are in a state of decay, and the huge kettles for boiling deeply coated with rust.

The climate of Cuba offers all the essentials, heat, moisture and organic matter, for the development of germ life in its most active form.

The great heat and moisture, so excellent for the development of infected wounds, and for the rapid decomposing of the heavy undergrowth cannot, I believe, be exceeded anywhere.

The frequent tropical showers, invariably followed by a hot steam, along with which germs seem to float; the consequent exposure of the men to that glaring heat and moisture, lowered the general tone of the system so that they were especially liable to attacks of miasmatic diseases (malarial and typhoid fevers and dysentery.)

Owing to the dense humidity, clothing does not dry so long as it remains on the person, but must be removed, a condition that was absolutely impossible for many days on the field before Santiago. To this alone, much of our sickness may be attributed.

Our new camp, pitched on the eminence of El Caney, about one and one-half miles from the village, overlooking the city and bay of Santiago, with its excellent water, shade, grass, and increased comforts, which were daily shipped from our transports, presented a scene far more conciliatory than had been witnessed about the Tenth Horse for many days.

MEDALS OF HONOR AND CERTIFICATES OF MERIT GRANTED TO COLORED SOLDIERS FOR DISTINGUISHED SERVICES IN THE CUBAN CAMPAIGN. OFFICIAL.

MEDALS OF HONOR.

Name.	Rank	Regiment.	Troop or Co.	Remarks.
Bell, Dennis	Pvt.	10th Cav.	Troop H.	For gallantry in action at Tayabacoa, Cuba, June 30, 1898.
Lee, Fitz	Pvt.	10th Cav.	Troop M.	
Tompkins, Wm. H.	Pvt.	10th Cav.	Troop M.	
Wanton, Geo. H.	Pvt.	10th Cav.	Troop M.	

CERTIFICATES OF MERIT.

Name.	Rank.	Regiment.	Troop or Co.	Remarks.
Bates, James	Pvt.	9th Cav.	Troop H.	
Crosby, Scott	Pvt.	24th Inf.	Comp. A.	
Davis, Edward	Pvt.	9th Cav.	Troop H.	
Elliott, J.	Sergt.	10th Cav.	Troop D.	
Fasit, Benjamin	Sergt.	10th Cav.	Troop E.	
Gaither, O.	Q.M. Sergt	10th Cav.	Troop E.	
Goff, G. W.	Sergt.	9th Cav.	Troop B.	
Graham, J.	Sergt.	10th Cav.	Troop E.	
Hagen, Abram	Corp.	24th Inf.	Comp. G.	
Herbert, H. T.	Corp.	10th Cav.	Troop E.	
Houston, Adam	1st Sergt.	10th Cav.	Troop C.	
Jackson, J.	1st Sergt.	9th Cav.	Troop C.	
Jackson, Elisha	Sergt.	9th Cav.	Troop H.	
Jackson, Peter	Corp.	24th Inf.	Comp. G.	
Jefferson, C. W.	1st Sergt.	9th Cav.	Troop B.	
McCoun, P.	1st Sergt.	10th Cav.	Troop E.	
Moore, Loney	Pvt.	24th Inf.	Comp. A.	
Oden, Oscar	Musician	10th Cav.	....	
Payne, William	Sergt.	10th Cav.	Troop E.	
Pumphrey, Geo. W	Corp.	9th Cav.	Troop H.	
Satchell, James	Sergt.	24th Inf.	Comp. A.	
Smith, L.	Pvt.	10th Cav.	Troop D.	
Thornton, William	Corp.	24th Inf.	Comp. G.	
Walker, J.	Corp.	10th Cav.	Troop D.	
Williams, John T.	Sergt.	24th Inf.	Comp. G.	
Williams, R.	Corp.	24th Inf.	Comp. B.	

Besides the Certificates of Merit and Medals of Honor, mentioned above, and the promotions to commissions in the volunteer services, there were some instances of promotion to non-commissioned officers' positions of men in the ranks or junior grade for conspicuous gallantry. Notably among such were Benjamin F. Sayre, of the Twenty-fourth, promoted to Sergeant-Major for gallantry at San Juan, and Private James W. Peniston, of the Tenth Cavalry, promoted to Squadron Sergeant-Major for conspicuous bravery at Las Guasimas. Others there may be whose names are not available at this time.

CHAPTER XI.

THE COLORED VOLUNTEERS.

The Ninth Ohio Battalion—Eighth Illinois—Twenty-third Kansas—Third North Carolina—Sixth Virginia—Third Alabama—The Immunes.

The return of the army and the repatriation of the Spanish army from Cuba, brought before the country for immediate solution the problem of garrisoning that island; and in a very short time the question of similar nature regarding Porto Rico. Ten regiments of immunes had been organized in the volunteer service partly in anticipation of such a situation. Four of these regiments were composed of colored enlisted men. The regiments were classed as United States Volunteer Infantry, and were numbered from one to ten, the Seventh, Eighth, Ninth and Tenth being colored.

Of these four colored regiments the officers above first lieutenants were white men, except the chaplains, and in some cases the surgeons. Very little care had been taken in enlisting the men, as it was important to get the regiments in the field as soon as possible; yet of them as a whole General Breckinridge, Inspector-General, speaks as follows: "The colored regiments of immunes, so called, raised for this war, have turned out, so far as can be judged from their camp life (as none of them have been in any actual campaign), very satisfactory. The regular colored regiments won golden opinions in battle. The experiment of having so many colored officers has not yet shown its full results. Certainly we should have the best obtainable officers for our volunteers, and therefore some such men as Colonel Young, who is a graduate of the Military Academy at West Point, whether white or black, must be sought for."

Besides these four colored regiments of immunes, so-called, there were other State organizations composed entirely of colored men, mustered into the United States service, as for example the Ninth Battalion of the Ohio National Guard. This organization was composed of four companies, with colored captains and lieutenants, the staff officers also being colored, the commanding officer of the battalion being Major Young, who was a first lieutenant in the Regular Army, a graduate from the Military Academy, and an officer of experience. He is the person referred to as *Colonel* Young by General Breckinridge, cited just above. This battalion, although not permitted to do any active campaigning, maintained itself well in that most trying of all duties for raw troops—camp duty—winning a good record in the South as well as in the North, having been stationed in Virginia, Pennsylvania and lastly in South Carolina; from which latter place it was mustered out, and the men proceeded to their homes in an orderly manner, reflecting credit upon themselves and the officers under whom they had served. This organization is mentioned first, because it was the only one of its kind commanded by a Regular Army officer, and a man who had received scientific military training.*

Two of these volunteer regiments, the Eighth Illinois and the Twenty-third Kansas, reached Cuba and made history there, in garrison service, coming in direct contact with the Ninth Immunes, and in no sense suffering in comparison there-

*See "Outline History of the Ninth (Separate) Battalion Ohio Volunteer Infantry," by the Battalion Adjutant, Lieutenant Nelson Ballard, following the close of this chapter.

to. The Eighth Illinois being the first to go to the front, in a sense deserves to be noticed here first. This remarkable regiment was developed out of the Ninth Battalion, Illinois National Guard, and owes its origin to the persistent efforts of Messrs. John R. Marshall, Robert R. Jackson, Franklin Dennison, E. H. Wright, Rev. R. C. Ransom, Rev. J. W. Thomas, S. B. Turner and doubtless many others whose names do not appear. These gentlemen named called upon the Governor of their State the next day after the President had issued his call for 175,000 volunteers, and received from that official the assurance that if another call should be made they should have the opportunity to recruit their battalion to a regiment, and that he would "call that regiment first into the service," and "that every officer in that regiment will be a colored man."

After receiving this encouragement, the leaders began at once the work of organizing and recruiting, and when the second call came, May 25th, the regiment was well under way, and soon ready to go into camp to prepare for service. On June 30th it assembled in Springfield from the following places: Seven hundred men from Chicago; one hundred and twenty from Cairo; a full company from Quincy, and smaller numbers from Mound City, Metropolis and Litchfield, and nearly a company from Springfield. The regiment was sworn in during the latter half of July, the muster roll showing 1,195 men and 46 officers, every one of whom was of African descent except one private in a Chicago company.

Of these forty-six officers, ten had received college education, six were lawyers, and the others were educated in the public schools, or had served in the Regular Army as non-commissioned officers. Many of them were directly from Illinois, that is in the sense of having been born and reared in the

State, and were fully accustomed to the full exercise of their rights as men and citizens. In character and intelligence the official element of the Eighth was about up to the standard of the volunteer army, as events subsequently proved.

Going into camp with the Ninth, white, this latter reigment, early in August, received an order to move to a Southern camp en route for Cuba, leaving the Eighth behind, greatly to the chagrin of both officers and men. Governor Tanner was evidently disturbed by this move, and expressed himself in the following language: "Even from the very doors of the White House have I received letters asking and advising me not to officer this regiment with colored men, but I promised to do so, and I have done it. I shall never rest until I see this regiment —my regiment—on the soil of Cuba, battling for the right and for its kinsmen."

Later the misfortunes of the First Illinois proved the opportunity of the Eighth. This regiment was in Cuba, suffering terribly with the fever, the men going down under its effects so rapidly that the Colonel in command implored Governor Tanner "to use all influence at Washington to secure the immediate recall of the First Illinois." When the Governor received this message he sent for Colonel Marshall, of the Eighth, and asked him to ascertain the sentiments of the officers and men of his regiment in regard to being sent to relieve the First. On the 4th day of August Colonel Marshall was able to send to Washington the following dispatch:

"H. C. Corbin, Adjutant-General:—

"I called the officers of the Eighth Illinois, colored, in conference and they are unanimously and enthusiastically in favor of being sent to relieve the First Illinois at Santiago."

To this hearty dispatch came the following reply:

"The Secretary of War appreciates very much the offer of the Eighth Illinois Volunteer Infantry for duty in Santiago, and has directed that the regiment be sent there by steamer Yale, leaving New York next Tuesday. The main trouble with our troops now in Cuba is that they are suffering from exhaustion and exposure incident to one of the most trying campaigns to which soldiers have ever been subjected."

"H. C. CORBIN,
"*Adjutant-General.*"

This action on the part of the regiment is said to have so pleased the President that on hearing it he declared it was the proudest moment of his life.

On the 9th of August the regiment left Springfield, and in passing through Illinois and Ohio was greeted with the most generous enthusiasm, the people supplying the men with free lunches at every station. This was the period when the sympathy of the whole country was turned toward the colored soldier in consequence of the reports of valor and heroism that had been circulated concerning the black regulars. On the afternoon of the 11th the Yale cast off her lines, and with the first American Negro regiment that the world has ever seen, steamed out of New York harbor amid the ringing of bells and shrieks of steam whistles, and four days later, August 15, landed in Cuba. The regiment remained in Cuba until March 10, perfoming garrison duty so well that General Breckenridge said it was "as fine a volunteer regiment as was ever mustered into the service," and that it was "a shame to muster out of service such an excellent regiment."

The Twenty-third Kansas, made up in that State and officered as was the Eighth Illinois, by men of the same race, with the enlisted men, arrived in Cuba August 30, and in company with the Eighth Illinois Regiment, was stationed in the country about San Luis, with headquarters at that place, Colonel Marshall, of the Illinois Regiment, serving as commander of the post, and also as Governor of the Province of San Luis,

A detachment of the Illinois Regiment, under command of Major Jackson, was sent to Palma Soriana, and did excellent work there in the preservation of order between the Cubans and Spaniards, who were living together in that place in outward peace but in secret resentful hostility. Major Jackson managed affairs so well that both parties came to admire him, and when he was called away expressed their regret. Captain Roots, who commanded the post after the departure of Major Jackson, was equally fortunate, especially with the Cubans, and when it was thought his command was to be removed, the citizens generally united in a petition to the General commanding, asking that both the Captain and his command might remain in the city. The fact is also noted by the chroniclers of the regiment that several marriages took place in Palma Soriana between soldiers of the Eighth Illinois and Cuban maidens.

The Eighth Regiment was finally settled in San Luis, occupying the old Spanish barracks and arsenal, and under Colonel Marshall's supervision the city was put in fine sanitary condition, streets and yards being carefully policed; meanwhile under the reign of order and peace which the Colonel's just methods established, confidence prevailed, business revived and the stagnation which had so long hung like a fog over the little city, departed, and in its stead came an era of bustling activity.

All was peaceful and prosperous, both with the citizens and the garrison, until the Ninth United States Volunteers came in the vicinity. Then a difficulty sprang up in which both regiments became involved, although it was in no sense serious, but it afforded a pretext for the removal of the Eighth Illinois from the city. The event turned out all the better for the

Eighth, as it enabled them to establish Camp Marshall, about three miles from the city, in a healthy neighborhood, where they remained until ordered home to be mustered out. The regiment came back to Chicago in fine condition and was tendered an enthusiastic welcome by that great city. Thus two entire regiments represented the country abroad in this, its first, foreign war with a European power.

It should also be recorded that although the Ninth United States Volunteers was composed of persons who were classed as immune, and had come chiefly from Louisiana, and notwithstanding that the officers of the regiment above lieutenants were white men, and the colonel an officer of the Regular Army of long experience, and was specially praised by so good a sanitarian as General Wood for having been constant and untiring in his efforts to look after the welfare of his men, and that the surgeons of the regiment were white men, that deaths among the colored men numbered one officer and seventy-three enlisted men. In striking contrast with this record of the immune regiment is that of the Eighth Illinois, which was made up entirely of residents of that State and officered throughout by colored men. Its medical officers were men of high character, and its losses by death were just twenty, or but little over one-fourth the number that occurred in the immune regiment. An efficient auxiliary society to this regiment was formed of colored ladies of Chicago who forwarded to the sick in Cuba more than six hundred dollars worth of well chosen supplies, which did much for the comfort of those in the hospital; but this would not account for the great difference in the death rate of the two regiments. Though not immune, the Eighth Illinois fared very much better than the so-called immune regiment, although the latter had the benefit of white officers. The experience of the Twenty-third Kansas did not differ in any important respect from that of the Eighth Illinois. Both regiments returned to their homes in March, 1899, and were mustered out of the service, leaving behind them good records for efficiency.

The Sixth Virginia Regiment consisted of eight companies and was under command of Lieutenant-Colonel Richard C. Croxton, of the Regular Army, white, with Majors J. B. Johnson and W. H. Johnson, colored. It was mustered into service during the latter part of the summer and went into camp near Knoxville, Tennessee. Here an order came from Corps Headquarters, at Lexington, Kentucky, directing that nine of the officers, including one major, should appear before a board of examiners in order to give evidence of their fitness to command. The officers named, regarding this as uncalled for, immediately tendered their resignations. The vacancies thus created were filled by the Governor of the State, the appointees being white men. These white officers on arriving at the camp and finding themselves unwelcome, immediately followed in the wake of their colored predecessors, and tendered their resignations.

The difficulties arising from this friction were somehow adjusted, but in what manner the reports available at this time do not show. Moving to Macon, Georgia, the regiment remained in the service until some time in the winter, when it was mustered out. Much was said by the local papers to the detriment of the men composing this regiment, but viewing their action from the standpoint of the civilian and citizen, it does not appear reprehensible. They had volunteered with the understanding that their own officers, officers with whom they were well acquainted, and in whose friendship they held a place, should command them, and when they saw these officers displaced and white strangers put in their stead, they felt a pardonable indignation, and took their own way of expressing it. As soldiers, their conduct in resisting authority, cannot be commended.

The Third North Carolina Volunteer Infantry was organized as were the regiments of Illinois and Kansas, above described. The officers of the North Carolina Regiment were all colored men of that State and were men of character and note. Its commanding officer, Colonel Young, had held responsible positions under both State and National Governments, had been editor of a paper and member of the State Legislature and Major in the State militia. In character, he was above reproach, being a strict teetotaler and not even using tobacco. The regiment made a good record, but did not see any active service.

A peculiar regiment was organized in Alabama, known as the Third Alabama Volunteer Infantry, in which the enlisted men were all colored and the officers all white. The regiment saw no service and attracted no attention outside of its immediate locality.

Two companies of colored men with colored captains were also mustered into the United States service from Indiana, and finally attached to Colonel Huggins' command, although not becoming a part of his regiment, the Eighth Immunes. They were stationed at Fort Thomas, Ky., and at Chickamauga, and were mustered out early. Their officers were men of intelligence who had acquired experience by several years' service in the militia, and the companies were exceptionally well drilled. They were designated Companies A and B and were commanded by Captains Porter and Buckner, with Lieutenant Thomas as Quartermaster.

The organization of the four immune regiments, already mentioned, gave opportunity for ninety-six colored men to obtain commissions as lieutenants. A few of these positions were seized upon by influential young white men, who held them, but with no intention of ever serving in the regiments, as they found staff positions much more congenial to their tastes. The colored men who were appointed lieutenants in these regiments were generally either young men of ability and influence who had assisted in getting up their companies, and who in many cases had received some elementary military instruction as cadets in school, or men who had distinguished themselves by efficiency or gallantry in the Regular Army. Some exceptions there were, of course, and a few received commissions in consequence of personal friendship and political considerations. Before these regiments were mustered out of the service about one-half of the lieutenants were men from the Regular Army.

I am sure the reader will be pleased to learn that Sergeants Foster, Buck and Givens, whose deeds in Cuba have already been related, were rewarded with commissions, and that the gallant Thomas C. Butler, who rushed forward from his company's line and seized the Spanish standard at El Caney, was afterward permitted to serve in Cuba with the rank of a commissioned officer. Besides those named above, there were others also of marked ability and very respectable attainments who received commissions on general merit, as well as for gallantry. Chief among the class promoted for efficiency was First Lieutenant James R. Gillespie, formerly Post Quartermaster-Sergeant. Gillespie had served several years in the Tenth Cavalry and had proved himself an excellent soldier. Both in horsemanship and as marksman he was up to the standard, while his character and business qualifications were such as to secure for him a staff position of responsibility. As Quartermaster-Sergeant he held positions of important posts and filled them with great satisfaction. Because of his efficiency as a soldier he was given a commission as First Lieutenant and executed the duties of his office with the same ability that had marked his career as an enlisted man. From the Tenth Cavalry also came First Lieutenant Baker, whose commission was a tribute to his fidelity and efficiency. A soldier of high type he bore his commission and its honors as worthily as any son of our Republic. In the same category must be placed First Lieutenant Wm. McBryer, formerly Sergeant in the Twenty-fifth Infantry. McBryer had served in the Tenth Cavalry and had won a Medal of Honor in conflict with the Indians. He was a soldier distinguished by strength of character, prompt executiveness, quick decision and courage. He was also possessed of considerable literary skill, was a good speaker and attractive writer, and a man of fine parts. He was a valuable acquisition to the volunteer service and would have made a fine captain.

Of the colored sergeants from the Regular Army who were given commissions in the volunteer service it would not be extravagant to say that all were men of worth, well-tried in the service, and there was scarcely one of them but could have successfully commanded a company. Lieutenant A. J. Smith, formerly First Sergeant in the Twenty-fifth Infantry, was so well informed in the paper work of the army and in company administration particularly that he was regarded as an authority, and he was so well experienced in the whole life of a soldier, in camp, field, garrison and in battle, that it would have been difficult to find his superior in the army. To the credit of all of the enlisted men of the Regular Army referred to, who received commissions in the volunteer service, all served honorably and were mustered out without bringing any scandal of any sort upon the service.

The colored volunteers in the service acquitted themselves as well as the average volunteer, and when mustered out proceeded to their homes about as others did. The treatment accorded them in some of the Southern cities, especially in Nashville, Tennessee, did not speak well for the loyalty of that section, nor was it such as might reasonably be expected from a people who had fared so well in the offices and honors of the short war. From the best sources available, it seems incumbent to say that the many charges alleged against the colored volunteers for excessive rioting and disorder were without proper foundation, and the assaults made upon them unjustifiable and cruel. The spirit of the assailants is best seen from a description of the attack made upon the unarmed discharged soldiers of the Eighth Immuners in Nashville, already alluded to. This description was made by the sheriff who participated in the brutality. An officer who was on the train, and who was asleep at the time, when aroused went into the car where the men were and found that they had been beaten and robbed, and in some instances their discharges taken from them and torn up, and their weapons and money taken from them by citizens. It was about one o'clock A. M. and the men were generally asleep when attacked. The sheriff gloats over it in language which ought not be allowed to disappear:

"It was the best piece of work I ever witnessed. The police went to the depot, not armed with the regulation 'billy,' but carrying stout hickory clubs about two and one-half feet long.

Their idea was that a mahogany or lignum vitae billy was too costly a weapon to be broken over a Negro's head. The police were on board the train before it stopped even, and the way they went for the Negroes was inspiring. The police tolerated no impudence, much less rowdyism, from the Negroes, and if a darky even looked mad, it was enough for some policeman to bend his club double over his head. In fact after the police finished with them they were the meekest, mildest, most polite set of colored men I ever saw." This language is respectfully dedicated to the memory of the proud city of Nashville, and presents to the readers the portrait of her police.

Despite this vile treatment, the colored soldier went on to his home, ready again to respond to his country's call, and to rally to the defence of his country's flag, and, incidentally, to the preservation of the lives and homes of the misguided, heartless beings who can delight in his sufferings. The hickory club belongs to one sort of warrior; the rifle to quite another. The club and rifle represent different grades of civilization. The Negro has left the club; the language from Nashville does honor to the club. Billy and bully are the theme of this officer of the law, and for a "darkey even to look mad" is ample justification for "some policeman to bend his club double over his head." Were these policemen rioters? Or were they conservaters of the peace? Judge ye!

OUTLINE HISTORY OF THE NINTH (SEPARATE) BATTALION, OHIO VOLUNTEER INFANTRY.

By the Battalion Adjutant, Lieutenant Wilson Ballard.

The Ninth Battalion, Ohio Volunteer Infantry, the only colored organization from Ohio in the Volunteer Army during the war with Spain, was, previous to the date of its muster into the United States service, known as the Ninth Battalion, Ohio National Guard. April 25th, 1898, the battalion, consisting of three companies, A from Springfield, under Captain R. R. Rudd; B from Columbus, under Captain James Hopkins, and C from Xenia, under Captain Harry H. Robinson, was ordered into camp at Columbus, Ohio. The battalion was under the command of Major Charles Fillmore.

May 14, 1898, the battalion was mustered into the volunteer service by Captain Rockefeller, U. S. A. Lieutenant Charles Young, U. S. A., then on duty at Wilberforce University, Wilberforce, Ohio, as professor of military science and tactics, was commissioned by Governor Bushnell as Major commanding the Ninth Battalion, O. V. I., relieving Major Fillmore. In order to enable Lieutenant Young to accept his volunteer commission, he was granted an indefinite leave of absence by the War Department.

May 19, 1898, the command having been ordered to join the Second Army Corps at Camp Russell A. Alger, near Falls Church, Va., left Camp Bushnell and arrived at Camp Alger May 21, 1898.

When Major-General Graham assumed command of the Second Army Corps and organized it into divisions, the battalion was placed in the provisional division. In June (exact date not remembered) the battalion was placed in the Second Brigade, Second Division, being brigaded with the Twelfth Pennsylvania and Seventh Illinois Regiments. The battalion was relieved from the Second Brigade, Second Division and placed in the Second Brigade, First Division, being brigaded with the Eighth Ohio and Sixth Massachusetts.

A New Jersey regiment was relieved from duty as corps

headquarters' guard late in June and the Ninth Battalion assigned to that duty. The battalion performed this duty until it was ordered South from Camp Meade, Penn., when it became separated from corps headquarters. Important outposts, such as the entrance to Falls Church and the guarding of the citizens' gardens and property, were under the charge of the command.

When General Garretson's brigade (Second Brigade, First Division, consisting of the Eighth Ohio, Ninth Battalion and Sixth Massachusetts) was ordered to Cuba, General Graham, thinking that his entire Army Corps would soon be ordered to active service, requested the War Department, as the battalion was his headquarters guard, to let the battalion remain with him. (See telegrams Gen. Graham's report to the Secretary of War.) General Graham's request being honored by the department, the battalion was deprived of this chance of seeing active service in foreign fields. The battalion was then attached to the Second Brigade, Second Division, under Brigadier-General Plummer, being brigaded with the First New Jersey, Sixty-fifth New York and Seventh Ohio.

In July the battalion was relieved from this brigade and attached directly to corps headquarters. When the Second Army Corps was ordered to Camp Meade, Penna., the battalion was one of the first to break camp, going with corps headquarters. The battalion left Camp R. A. Alger August 15, 1898, and arrived in camp at Camp George G. Meade, near Middletown, Penna., August 16, 1898. In camp the battalion occupied a position with the signal and engineer corps and hospital, near corps headquarters.

When the Peace Jubilee was held in Philadelphia, the battalion was one of the representative commands from the Second Army Corps, being given the place of honor in the corps in the parade, following immediately General Graham and staff. When the corps was ordered South the battalion was assigned to the Second Brigade under Brigadier-General Ames. The battalion left Camp Meade November 17. Up to this time it had done the guard duty of corps headquarters and was complimented for its efficient work by the commanding general. The battalion arrived in Summerville, S. C., November 21, 1898. It was brigaded with the Fourteenth Pennsylvania and Third Connecticut.

When the battalion arrived in the South the white citizens were not at all favorably disposed toward colored soldiers, and it must be said that the reception was not cordial. But by their orderly conduct and soldierly behavior the men soon won the respect of all, and the battalion was well treated before it left. November 28-29 Major Philip Reade, Inspector General First Division, Second Army Corps, inspected the Ninth Battalion, beginning his duties in that brigade with this inspection. He complimented the battalion for its work both from a practical and theoretical standpoint. Coming to the Fourteenth Pennsylvania he required them to go through certain movements in the extended order drill which not being done entirely to his satisfaction, he sent his orderly to the commanding officer of the Ninth Battalion, requesting him to have his command on the drill ground at once. The battalion fell in and marched to the ground and when presented to the Inspector orders were given for it to go through with certain movements in the extended order drill in the presence of the Pennsylvania regiment. This done, the Inspector dismissed the battalion, highly complimenting Major Young on the efficiency of his command. Just after the visit of the Inspector General, General S. B. M. Young, commanding the Second Army Corps, visited Camp Marion. Orders were sent to Major Young one morning to have his battalion fall in at once, as the General desired to have them drill. By his command the battalion went through the setting-up exercises and battalion drill in close and extended order. The General was so well pleased with the drill that the battalion was exempted from all work during the remainder of the day.

The battalion was ordered to be mustered out January 29, 1899. Lieutenant Geo. W. Van Deusen, First Artillery, who was detailed to muster out the command, hardly spent fifteen minutes in the camp. Major Young had been detailed Assistant Commissary of Musters and signed all discharges for the Ninth Battalion, except for the field and staff, which were signed by Lieutenant Van Deusen. The companies left for their respective cities the same night they were paid. Major Bullis was the paymaster.

CHAPTER XII.

COLORED OFFICERS.

By Captain Frank R. Steward, A. B., LL. B., Harvard, Forty-ninth U. S. Volunteer Infantry—Appendix.

Of all the avenues open to American citizenship the commissioned ranks of the army and navy have been the stubbornest to yield to the newly enfranchised. Colored men have filled almost every kind of public office or trust save the Chief Magistracy. They have been members of both Houses of Congress, and are employed in all the executive branches of the Government, but no Negro has as yet succeeded in invading the commissioned force of the navy, and his advance in the army has been exceedingly slight. Since the war, as has been related, but three Negroes have been graduated from the National Military Academy at West Point; of these one was speedily crowded out of the service; another reached the grade of First Lieutenant and died untimely; the third, First Lieutenant Charles Young, late Major of the 9th Ohio Battalion, U. S. Volunteers, together with four colored Chaplains, constitute the sole colored commissioned force of our Regular Army.

Although Negroes fought in large numbers in both the Revolution and the War of 1812, there is no instance of any Negro attaining or exercising the rank of commissioned officer. It is a curious bit of history, however, that in the Civil War those who were fighting to keep colored men enslaved were the first to commission colored officers. In Louisiana but a few days after the outbreak of the war, the free colored population of New Orleans organized a military organization, called the "Native Guard," which was accepted into the service of the State and its officers were duly commissioned by the Governor.*

These Negro soldiers were the first to welcome General Butler when he entered New Orleans, and the fact of the organization of the "Native Guard" by the Confederates was used by General Butler as the basis for the organization of three colored regiments of "Native Guards," all the line officers of which were colored men. Governor Pinchback, who was a captain in one of these regiments, tells the fate of these early colored officers.

---

*Headquarters Department of the Gulf,
New Orleans, August 22, 1862.

General Orders No. 63.

"Whereas, on the 23d day of April, in the year eighteen hundred and sixty-one, at a public meeting of the free colored population of the city of New Orleans, a military organization, known as the 'Native Guards' (colored), had its existence, which military organization was duly and legally enrolled as a part of the militia of the State, its officers being commissioned by Thomas O. Moore, Governor and Commander-in-Chief of the militia of the State of Louisiana, in the form following, that is to say:

The State of Louisiana.
(Seal of the State.)

By Thomas Overton Moore, Governor of the State of Louisiana, and commander-in-chief of the militia thereof.

" 'In the name and by the authority of the State of Louisiana: Know ye that —— ——, having been duly and legally elected captain of the "Native Guards" (colored), first division of the Militia of Louisiana, to serve for the term of the war,

" 'I do hereby appoint and commission him captain as aforesaid, to take rank as such, from the 2d day of May, eighteen hundred and sixty-one.

" 'He is, therefore, carefully and diligently to discharge the duties of his office by doing and performing all manner of things thereto belonging. And I do strictly charge and require all officers, non-commissioned officers and privates under his command to be obedient to his orders as captain; and he is to observe and follow such orders and directions, from time to time, as he shall receive from me, or the future Governor

"There were," he writes, "in New Orleans some colored soldiers known as 'Native Guards' before the arrival of the Federal soldiers, but I do not know much about them. It was a knowledge of this fact that induced General Butler, then in command of the Department of the Gulf, to organize three regiments of colored soldiers, viz: The First, Second and Third Regiments of Native Guards.

"The First Regiment of Louisiana Native Guards, Colonel Stafford commanding, with all the field officers white, and a full complement of line officers (30) colored, was mustered into service at New Orleans September 27, 1862, for three years. Soon after General Banks took command of the department and changed the designation of the regiment to First Infantry, Corps d'Afrique. April 4th, 1864, it was changed again to Seventy-third United States Colored Infantry.

---

of the State of Louisiana, or other superior officers, according to the Rules and Articles of War, and in conformity to law.

" 'In testimony whereof, I have caused these letters to be made patent, and the seal of the State to be hereunto annexed.

" 'Given under my hand, at the city of Baton Rouge, on the second day of May, in the year of our Lord one thousand eight hundred and sixty-one.

(L. S.) (Signed) THOS. O. MOORE.

" 'By the Governor:

(Signed) P. D. HARDY,
Secretary of State.

(Wilson: Black Phalanx, p. 194.)

*"On the 23d of November, 1861, there was a grand review of the Confederate troops stationed at New Orleans. An Associated Press despatch announced that the line was seven miles long. The feature of the review, however, was one regiment of fourteen hundred free colored men. Another grand review followed the next spring, and on the appearance of rebel negroes a local paper made the following comment:

" 'We must also pay a deserved compliment to the companies of free colored men, all very well drilled and comfortably uniformed. Most of these companies, quite unaided by the administration, have supplied themselves with arms without regard to cost or trouble. On the same day one of these negro companies was presented with a flag, and every evidence of public approbation was manifest.' "

(Williams's Negro Troops in the Rebellion, pp. 83-4.)

"The Second Louisiana Native Guards, with Colonel N. W. Daniels and Lieutenant-Colonel Hall, white, and Major Francis E. Dumas, colored, and all the line officers colored except one Second Lieutenant, was mustered into service for three years, October 12, 1862. General Banks changed its designation to Second Infantry Corps d' Afrique, June 6, 1863, and April 6, 1864, it was changed to Second United States Colored Troops. Finally it was consolidated with the Ninety-first as the Seventy-fourth Colored Infantry, and mustered out October 11, 1865.

"The Third Regiment of Louisiana Native Guards, with Colonel Nelson and all field officers white, and all line officers (30) colored, was mustered into service at New Orleans for three years, November 24, 1862. Its designation went through the same changes as the others at the same dates, and it was mustered out November 25, 1865, as the Seventy-fifth Colored Infantry.

"Soon after the organization of the Third Regiment, trouble for the colored officers began, and the department began a systematic effort to get rid of them. A board of examiners was appointed and all COLORED officers of the Third Regiment were ordered before it. They refused to obey the order and tendered their resignations in a body. The resignations were accepted and that was the beginning of the end. Like action with the same results followed in the First and Second Regiments, and colored officers were soon seen no more. All were driven out of the service except three or four who were never ordered to appear before the examining board. Among these was your humble servant. I was then Captain of Company A, Second Regiment, but I soon tired of my isolation and resigned."

Later on in the war, with the general enlistment of colored soldiers, a number of colored chaplains and some surgeons were commissioned. Towards the close of the war several colored line officers and a field officer or two were appointed. The State of Massachusetts was foremost in according this recognition to colored soldiers. But these later appointments came, in most cases, after the fighting was all over, and gave few opportunities to command. At the close of the war, with the muster out of troops, the colored officers disappeared and upon the reorganization of the army, despite the brilliant record of the colored soldiers, no Negro was given a commission of any sort.

The outbreak of the Spanish War brought the question of colored officers prominently to the front. The colored people began at once to demand that officers of their own race be commissioned to command colored volunteers. They were not to be deluded by any extravagant praise of their past heroic services, which veiled a determination to ignore their just claims. So firmly did they adhere to their demands that but one volunteer regiment of colored troops, the Third Alabama, could be induced to enter the service with none of its officers colored. But the concessions obtained were always at the expense of continuous and persistent effort, and in the teeth of a very active and at times extremely violent opposition. We know already the kind of opposition the Eighth Illinois, the Twenty-third Kansas, and the Third North Carolina Regiments, officered entirely by colored men, encountered. It was this opposition, as we have seen, which confined colored officers to positions below the grade of captain in the four immune regiments. From a like cause, we know also, distinguished non-commissioned officers of the four regular regiments of colored troops were allowed promotion only to Lieutenantcies in the immune regiments, and upon the muster out of those organizations, were compelled, if they desired to continue soldiering, to resume their places as enlisted men.

There is some explanation for this opposition in the nature of the distinction which military rank confers. Military rank and naval rank constitute the only real distinction among us. Our officers of the army and navy, and of the army more than of the navy, because the former officers are more constantly within the country, make up the sole separate class of our population. We have no established nobility. Wealth confers no privilege which men are bound to observe. The respect paid to men who attain eminence in science and learning goes only as far as they are kown. The titles of the professions are matters of courtesy and customs only. Our judges and legislators, our governors and mayors, are still our "fellow citizens," and the dignity they enjoy is but an honorary one. The highest office within our gift offers no exception. At the close of his term, even an ex-President, "that melancholy product of our system," must resume his place among his fellow citizens, to sink, not infrequently, into obscurity. But fifty thousand soldiers must stand attention to the merest second lieutenant! His rank is a *fact*. The life tenure, the necessities of military discipline and administration, weld army officers into a distinct class and make our military system the sole but necessary relic of personal government. Any class with special privileges is necessarily conservative.

The intimate association of "officer" and "gentleman," a legacy of feudal days, is not without significance. An officer must also be a gentleman, and "conduct unbecoming an officer and a gentleman" is erected into an offence punishable by dismissal from the service. The word "gentleman" has got far away from the strict significance of its French parent. De Tocqueville has made us see the process of this development. Passing over to England, with the changing conditions, "gentleman" was used to describe persons lower and lower in the social scale, until, when it crossed to this country, its significance became lost in an indiscriminate application to all citizens.* A flavor of its caste significance still remains in the traditional "high sense of honor" characteristic of our military service. It was a distant step for a slave and freedman to become an officer and gentleman.

While the above reflections may be some explanations *in fact* for the opposition to the commissioning of Negroes, there was no one with hardihood enough to bring them forward. Such notions might form the groundwork of a prejudice, but they could not become the reason of a policy. It is an instinctive tribute to the good sense of the American people that the opponents of colored officers were compelled to find reasons of another kind for their antagonism.

The one formula heard always in the campaign against colored officers was: Negroes cannot command. This formula was sent forth with every kind of variation, from the fierce fulminations of the hostile Southern press, to the more apologetic and philosophical discussions of our Northern secular and religious journals. To be sure, every now and then, there were exhibitions of impatience against the doctrine. Not a few newspapers had little tolerance for the nonsense. Some former commanders of Negro soldiers in the Civil War, notably, General T. J. Morgan, spoke out in their behalf. The

*De Tocqueville: L'Ancien Régime et La Revolution, p. 125-6.

brilliant career of the black regulars in Cuba broke the spell for a time, but the re-action speedily set in. In short it became fastened pretty completely in the popular mind as a bit of demonstrated truth that Negroes could not make officers; that colored soldiers would neither follow nor obey officers of their own race.

This formula had of course to ignore an entire epoch of history. It could take no account of that lurid program wrought in the Antilles a century ago—a rising mob of rebel slaves, transformed into an invincible army of tumultuous blacks, under the guidance of the immortal Toussaint, overcoming the trained armies of three Continental powers, Spain, England and France, and audaciously projecting a black republic into the family of nations, a program at once a marvel and a terror to the civilized world.

Not alone in Hayti, but throughout the States of Central and South America have Negroes exercised military command, both in the struggles of these states for independence, and in their national armies established after independence. At least one soldier of Negro blood, General Dumas, father of the great novelist, arose to the rank of General of Division in the French Army and served under Napoleon. In our day we have seen General Dodds, another soldier of Negro blood, returning from a successful campaign in Africa, acclaimed throughout France, his immense popularity threatening Paris with a renewal of the hysterical days of Boulanger. Finally, we need not be told that at the very head and front of the Cuban Rebellion were Negroes of every hue, exercising every kind of command up to the very highest. We need but recall the lamented Maceo, the Negro chieftain, whose tragic end brought sorrow and dismay to all of Cuba. With an army thronging with blacks and mulattoes, these Cuban chieftains, black, mulatto and white, prolonged such an harassing warfare as to compel the intervention of the United States. At the end of this recital, which could well have been extended with greater particularity, if it were thought needful, we are bound to conclude that the arbitrary formula relied upon by the opponents of colored officers was never constructed to fit such an obstinate set of facts.

The prolonged struggle which culminated in permitting the Negro's general enlistment in our Civil War had only to be repeated to secure for him the full pay of a soldier, the right to be treated as a prisoner of war, and to relieve him of the monopoly of fatigue and garrison duty. He was too overjoyed with the boon of fighting for the liberation of his race to make much contention about who was to lead him. With meagre exception, his exclusive business in that war was to carry a gun. Yet repeatedly Negro soldiers evinced high capacity for command. Colonel Thomas Wentworth Higginson draws a glowing portrait of Sergeant Prince Rivers, Color-Sergeant of the First South Carolina Volunteers, a regiment of slaves, organized late in 1862. The Color-Sergeant was provost-Sergeant also, and had entire charge of the prisoners and of the daily policing of the camp.

"He is a man of distinguished appearance and in old times was the crack coachman of Beaufort. * * * They tell me that he was once allowed to present a petition to the Governor of South Carolina in behalf of slaves, for the redress of certain grievances, and that a placard, offering two thousand dollars for his re-capture is still to be seen by the wayside between here and Charleston. He was a sergeant in the old 'Hunter Regiment,' and was taken by General Hunter to New

York last spring, where the chevrons on his arm brought a mob upon him in Broadway, whom he kept off till the police interfered. There is not a white officer in this regiment who has more administrative ability, or more absolute authority over the men; they do not love him, but his mere presence has controlling power over them. He writes well enough to prepare for me a daily report of his duties in the camp; if his education reached a higher point I see no reason why he should not command the Army of the Potomac. He is jet-black, or rather, I should say, wine-black, his complexion, like that of others of my darkest men, having a sort of rich, clear depth, without a trace of sootiness, and to my eye very handsome. His features are tolerably regular, and full of command, and his figure superior to that of any of our white officers, being six feet high, perfectly proportioned, and of apparently inexhaustable strength and activity. His gait is like a panther's; I never saw such a tread. No anti-slavery novel has described a man of such marked ability. He makes Toussaint perfectly intelligible, and if there should ever be a black monarchy in South Carolina he will be its king."*

Excepting the Louisiana Native Guards, the First South Carolina Volunteers was the first regiment of colored troops to be mustered into the service in the Civil War. The regiment was made up entirely of slaves, with scarcely a mulatto among them. The first day of freedom for these men was passed in uniform and with a gun. Among these Negroes, just wrested from slavery, their scholarly commander, Colonel Higginson, could find many whom he judged well fitted by nature to command.

"Afterwards I had excellent battalion drills," he writes,

---

*Thomas Westworth Higginson: Army Life in a Black Regiment, pp. 57-8.

"without a single white officer, by way of experiment, putting each company under a sergeant, and going through the most difficult movements, such as division columns and oblique squares. And as to actual discipline, it is doing no injustice to the line-officers of the regiment to say that none of them received from the men more implicit obedience than Color-Sergeant Rivers. * * * It always seemed to me an insult to those brave men to have novices put over their heads, on the ground of color alone, and the men felt it the more keenly as they remained longer in the service. There were more than seven hundred enlisted men in the regiment, when mustered out after more than three years' service. The ranks had been kept full by enlistment, but there were only fourteen line-officers instead of the full thirty. The men who should have filled these vacancies were doing duty as sergeants in the ranks."*

Numerous expeditions were constantly on foot in the Department of the South, having for their object the liberation of slaves still held to service in neighborhoods remote from the Union camps, or to capture supplies and munitions of war. Frequently these expeditions came in conflict with armed bodies of rebels and hot engagements would ensue, resulting in considerable loss of life. Colored soldiers were particularly serviceable for this work because of their intimate knowledge of the country and their zeal for the rescue of their enslaved brethren.

One of these expeditions, composed of thirty colored soldiers and scouts, commanded by Sergeant-Major Henry James, Third United States Colored Troops, left Jacksonville, Florida, early in March, 1865, to penetrate into the interior

*Tho[illegible] [illegible]ntworth Higginson: Army Life in a Black Regiment, p. 261.

through Marion county. They destroyed considerable property in the use of the rebel government, burned the bridge across the Oclawaha River, and started on their return with ninety-one Negroes whom they had rescued from slavery, four white prisoners, some wagons and a large number of horses and mules. They were attacked by a rebel band of more than fifty cavalry. The colored soldiers commanded by one of their own number, defeated and drove off the rebels, inflicting upon them the heavy loss of thirty men. After a long and rapid march they arrived at St. Augustine, Florida, with a loss of but two killed and four wounded, the expedition covering in all five days. These colored soldiers and their colored commander were thanked in orders by Major-General Q. A. Gilmore, commanding the department, who was moved to declare that "this expedition, planned and executed by colored men, under the command of a colored non-commissioned officer, reflects credit upon the brave participants and their leader," and "he holds up their conduct to their comrades in arms as an example worthy of emulation."*

It was no uncommon occurrence throughout the Civil War for colored non-commissioned officers to be thrown into command of their companies by the killing or wounding of their superior officers. On many a field of battle this happened and these colored non-commissioned officers showed the same ability to take the initiative and accept the responsibility, and conducted their commands just as bravely and unfalteringly as did their successors on the firing line at La Guasima and El Caney, or in the charge up San Juan Hill.

In the battle of New Market Heights, fought on the 29th of September, 1864, as part of a comprehensive effort to turn

*Williams's Negro Troops in the Rebellion, pp. 339-40, quoting the order.

Lee's left flank, the great heroism of the black soldiers, and the terrible slaughter among them, impressed their commander, the late Major-General Butler, to his dying day, and made him the stout champion of their rights for the rest of his life. In that battle, to quote from the orders putting on record the "gallant deeds of the officers and soldiers of the Army of the James":—

"Milton M. Holland, Sergeant-Major Fifth United States Colored Troops, commanding Company C; James H. Bronson, First Sergeant, commanding Company D; Robert Pinn, First Sergeant, commanding Company I, wounded; Powhatan Beaty, First Sergeant, commanding Company G, Fifth United States Colored Troops—all these gallant colored soldiers were left in command, all their company officers being killed or wounded, and led them gallantly and meritoriously through the day. For these services they have most honorable mention, and the commanding general will cause a special medal to be struck in honor of these gallant soldiers."

"First Sergeant Edward Ratcliff, Company C, Thirty-eighth United States Colored Troops, thrown into command of his company by the death of the officer commanding, was the first enlisted man in the enemy's works, leading his company with great gallantry for which he has a medal."

"Sergeant Samuel Gilchrist, Company K, Thirty-sixth United States Colored Troops, showed great bravery and gallantry in commanding his company after his officers were killed. He has a medal for gallantry."*

"Honorable mention" and "medals" were the sole reward open to the brave Negro soldiers of that day.

Not alone in camp and garrison, in charge of expeditions,

*Williams's Negro Troops in the Rebellion, pp. 334-6, original order quoted.

or as non-commissioned officers thrown into command of their companies on the field of battle have Negro soldiers displayed unquestioned capacity for command, but as commissioned officers they commanded in camp and in battle, showing marked efficiency and conspicuous gallantry. The colored officers of the First and Second Regiments of Louisiana Native Guards, whose history has been detailed earlier in this chapter,* were retained in the service long enough to command their troops in bloody combat with the enemy. It will be remembered that of the Second Regiment of Louisiana Native Guards only the Colonel and Lieutenant-Colonel were white, the Major, F. E. Dumas, and all the line officers, as in the case of the First Regiment of Louisiana Native Guards, being colored. On April 9, 1863, Colonel N. U. Daniels, who commanded the Second Regiment of Louisiana Native Guards, with a detachment of two hundred men of his regiment, under their colored officers, engaged and repulsed a considerable body of rebel infantry and cavalry at Pascagoula, Mississippi. The engagement lasted from 10 A. M. until 2 P. M. and was remarkable for the steadiness, tenacity and bravery of these black troops in this, their first battle, where they succeeded in defeating and beating off an enemy five times their number. The official report by the Colonel commanding declared: "Great credit is due to the troops engaged for their unflinching bravery and steadiness under this, their first fire, exchanging volley after volley with the coolness of veterans, and for their determined tenacity in maintaining their position, and taking advantage of every success that their courage and valor gave them; and also to their officers, who were cool and determined throughout the action, fighting their commands

*See pp. 351-6 MS.

against five times their number, and confident throughout of success. * * *

"I would particularly call the attention of the department to Major F. E. Dumas, Capt. Villeverd and Lieuts. Jones and Martin, who were constantly in the thickest of the fight, and by their unflinching bravery and admirable handling of their commands, contributed to the success of the attack, and reflected great honor upon the flag for which they so nobly struggled."*

The battle which settled for all time the bravery of black troops, and ought as well to silence all question about the capacity of colored officers, was the storming of Port Hudson, May 27, 1863. For months the Confederates had had uninterrupted opportunity to strengthen their works at Port Hudson at a time when an abundance of slave labor was at their disposal. They had constructed defenses of remarkable strength. On a bluff, eighty feet above the river, was a series of batteries mounting in all twenty siege guns. For land defenses they had a continuous line of parapet of strong profile, beginning at a point on the river a mile from Port Hudson and extending in a semi-circle for three or four miles over a country for the most part rough and broken, and ending again at the river, a half mile north of Port Hudson. At appropriate positions along this line four bastion works were constructed and thirty pieces of field artillery were posted. The average thickness of the parapet was twenty feet, and the depth of the ditch below the top of the parapet was fifteen feet. The ground behind the parapet was well adapted for the prompt movement of troops.*

---

*Wilson: Black Phalanx, p. 211, original order quoted.

---

*Campaigns of the Civil War. F. V. Greene. The Mississippi, p. 226 et seq.

On the 24th of May General Banks reached the immediate vicinity of Port Hudson, and proceeded at once to invest the place.

On the 27th the assault was ordered. Two colored regiments of Louisiana Native Guards, the First Regiment with all line officers colored, and the Third with white officers throughout, were put under command of Colonel John A. Nelson, of the Third Regiment, and assigned to position on the right of the line, where the assault was begun. The right began the assault in the morning; for some reason the left did not assault until late in the afternoon. Six companies of the First Louisiana and nine companies of the Third, in all 1080 men, were formed in column of attack. Even now, one cannot contemplate unmoved the desperate valor of these black troops and the terrible slaughter among them as they were sent to their impossible task that day in May. Moving forward in double quick time the column emerged from the woods, and passing over the plain strewn with felled trees and entangled brushwood, plunged into a fury of shot and shell as they charged for the batteries on the rebel left. Again and again that unsupported column of black troops held to their hopeless mission by the unrelenting order of the brigade commander, hurled itself literally into the jaws of death, many meeting horrible destruction actually at the cannon's mouth.

It was a day prodigal with deeds of fanatical bravery. The colors of the First Louisiana, torn and shivered in that fearful hail of fire, were still borne forward in front of the works by the color-sergeant, until a shell from the enemy cut the flag in two and gave the sergeant his mortal wound. He fell spattering the flag with blood and brains and hugged it to his bosom as he lay in the grasp of death. Two corporals sprang forward to seize the colors, contending in generous rivalry until a rebel sharpshooter felled one of them across the sergeant's lifeless body. The other dashed proudly forward with the flag. Sixteen men fell that day defending the colors.

Black officers and white officers commanded side by side, moving among the men to prompt their valor by word and example, revealing no difference in their equal contempt of death. Captain Quinn, of the Third Regiment, with forty reckless followers, bearing their rifles and cartridge boxes above their heads, swam the ditch and leaped among the guns, when they were ordered back to escape a regiment of rebels hastening for their rear. Six of them re-crossed alive, and of these only two were unhurt, the brave Quinn and a Lieutenant. The gallant Captain Andre Cailloux, who commanded the color company of the First Louisiana, a man black as night, but a leader by birth and education, moved in eager zeal among his men, cheering them on by words and his own noble example, with his left arm already shattered, proudly refusing to leave the field. In a last effort of heroism, he sprang to the front of his company, commanded his men to follow him, and in the face of that murderous fire, gallantly led them forward until a shell smote him to death but fifty yards from the works.

Cailloux, a pure Negro in blood, was born a freeman and numbered generations of freemen among his ancestry. He had fine presence, was a man of culture and possessed wealth. He had raised his company by his own efforts, and attached them to him, not only by his ardent pride of race, which made him boast his blackness, but also by his undoubted talents for command. His heroic death was mourned by thousands of his race who had known him. His body, recovered after the surrender, was given a soldier's burial in his own native city of New Orleans.

When the day was spent, the bleeding and shattered column was at length recalled. The black troops did not take the guns, but the day's work had won for them a fame that cannot die. The nation, which had received them into the service half-heartedly, and out of necessity, was that day made to witness a monotony of gallantry and heroism that compelled everywhere awe and admiration. Black soldiers, and led by black officers as well as white, assigned a task hopeless and impossible at the start, had plunged into that withering storm of shot and shell, poured fourth by artillery and infantry, charging over a field strewn with obstacles, and in madness of bravery had more than once thrown the thin head of their column to the very edge of the guns. They recoiled only to reform their broken lines and to start again their desperate work. When the day was gone, and they were called back, the shattered remnant of the column which had gone forth in the morning still burned with passion. With that day's work of black soldiers under black officers, a part forever of the military glory of the Republic, there are those who yet dare to declare that Negroes cannot command.

The assault on Port Hudson had been unsuccessful all along the line. A second assault was ordered June 13. It, too, was unsuccessful. The fall of Vicksburg brought the garrison to terms. The surrender took place July 9, 1863. In the report of the general commanding, the colored soldiers were given unstinted praise. General Banks declared that "no troops could be more determined or more daring."* The Northern press described glowingly their part in the fight.

*Williams's Negro Troops in the Rebellion, p. 221, original order quoted.

The prowess of the black soldiers had conquered military prejudice, and won for them a place in the army of the Union. And the brave black officers who led these black soldiers, they were, all of them, ordered forthwith before an examining board with the purpose of driving them from the service, and every one of them in self-respect was made to resign. In such manner was their bravery rewarded.

In the four regiments of colored troops made a part of the Regular Army since the Civil War, colored soldiers, to say nothing of the three colored graduates from West Point, referred to earlier in this chapter, have repeatedly given evidence of their capacity to command. An earlier chapter has already set forth the gallant manner in which colored non-commissioned officers, left in command by the killing or wounding of their officers, commanded their companies at La Guasima, El Caney and in the charge at San Juan. On numerous occasions, with none of the heroic setting of the Santiago campaign, have colored soldiers time and again command detachments and companies on dangerous scouting expeditions, and in skirmishes and fights with hostile Indians and marauders. The entire Western country is a witness of their prowess. This meritorious work, done in remote regions, has seldom come to public notice; the medal which the soldier wears, and the official entry in company and regimental record are in most cases the sole chronicle. A typical instance is furnished in the career of Sergeant Richard Anderson, late of the Ninth Cavalry. The sergeant has long ago completed his thirty years of service. He passed through all non-commissioned grades in his troop and regiment, and was retired as Post Commissary-Sergeant. The story of the engagements in which he commanded give ample proof of his ability and bravery. It would be no service to the sergeant to disturb his own frank and formal narrative.

The Sergeant's story:—

"While in sub-camp at Fort Cumming, New Mexico, awaiting orders for campaign duty against hostile Indians (old Naney's band), on the evening of June 5, 1880, my troop commander being absent at Fort Bayard, which left me in command of my troop, there being no other commissioned officer available, a report having come in to the commanding officer about 1 o'clock that a band of Apache Indians were marching toward Cook's Canon, Troops B and L, under general command of Captain Francis, 9th Cavalry, and myself commanding Troop B, were ordered out.

We came upon the Indians in Cook's Canon and had an engagement which lasted two or three hours. Three or four Indians were killed and several wounded. We had no men killed, but a few wounded in both L and B Troops. We followed the Indians many miles that evening, but having no rations, returned to Fort Cumming late that evening, and went into camp until the following morning, when the two troops took the trail and followed it many days, but being unable to overtake the Indians, returned to Fort Cumming.

In August, 1881, while my troop was in camp at Fort Cumming, New Mexico, awaiting orders for another campaign against these same Apache Indians, my troop commander having been ordered to Fort Bayard, New Mexico, on general court-martial duty, and during his absence having no commissioned officer available, I was in command of my troop subject to the orders of the post commander. At 12 o'clock at night, August 17, 1881, while in my tent asleep, the commanding officer's orderly knocked on my tent and informed me that the commanding officer wanted me to report to him at once. I asked the orderly what was up. He informed me that he supposed a scout was going out, as the commanding officer had sent for Lieutenant Smith, then in command of Troop H, 9th Cavalry.

I dressed myself promptly and reported, and found Lieutenant Smith and the commanding officer at the office on my arrival.

The commanding officer asked me about how many men I could mount for thirty days' detached duty, leaving so many men to take care of property and horses. I told him about how many. He ordered me to make a ration return for that number of men, and send a sergeant to draw rations for thirty days' scout; and for me to hurry up, and when ready to report to Lieutenant Smith. By 12.45 my troop was ready and mounted, and reported as ordered, and at 1 o'clock Troop's B and H pulled out from Fort Cumming for Lake Valley, New Mexico; and when the sun showed himself over the tops of the mountains we marched down the mountains into Lake Valley, thirty-five miles from Fort Cumming. We went into camp hoping to spend a few hours and take a rest, and feed our horses and men.

About 9 o'clock a small boy came running through camp crying as if to break his heart, saying that the Indians had killed his mother and their baby. Some of the men said the boy must be crazy; but many of them made for their horses without orders. Soon Lieutenant Smith ordered "Saddle up." In less than five minutes all the command was saddled up and ready to mount. We mounted and pulled out at a gallop, and continued at that gait until we came to a high mountain, when we came down to a walk. And when over the mountain we took up the gallop, and from that time on, nothing but a gallop and a trot, when the country was favorable for such. When we had marched about two miles from Lake Valley we met the father of the boy, with his leg bleeding where the Indians had shot him. We marched about half a mile farther, when we could see the Indians leaving this man's ranch. We had a running fight with them from that time until about 5 o'clock that evening, August 18th, 1881. Having no rations, we returned to Lake Valley with the intention of resting that night and taking the trail the next morning; but about 9 o'clock that night a ranchman came into camp and reported that the Indians had marched into a milk ranch and burned up the ranch, and had gone into camp near by.

Lieutenant Smith ordered me to have the command in readiness to march at 12 o'clock sharp, and said we could surprise those Indians and capture many of them and kill a few also. I went and made my detail as ordered, with five days' rations in haversacks, and at 12 o'clock reported as ordered.

About half-past 12 o'clock the command pulled out and marched within about a mile and a half of the milk ranch and went into camp; and at daylight in the morning saddled up and marched to the ranch. The Indians had pulled out a few minutes before our arrival. We took their trail and came up with them about 10 o'clock, finding the Indians in ambush. Lieutenant Smith was the first man killed, and when I heard his last command, which was "Dismount," then the whole command fell upon your humble servant. We fell back, up a canon and on a hill, and held them until 4 o'clock, when a reinforcement came up of about twenty men from Lake Valey and the Indians pulled off over the mountains. The following-named men were killed in the engagement:

Lieutenant G. W. Smith; Mr. Daily, a miner; Saddler Thomas Golding; Privates James Brown and Monroe Overstreet. Wounded—Privates Wesley Harris, John W. Williams and William A. Hallins.

After the Indians ceased firing and fell back over the mountains I cared for the wounded and sent Lieutenant Smith's body to Fort Bayard, New Mexico, where his wife was, which was about sixty miles from the battle-ground, and Mr. Daily's body to Lake Valley, all under a strong detachment of men under a non-commissioned officer; when I marched with the remainder of the command with the dead and wounded for Rodman Mill, where I arrived about 5 o'clock on the morning of August 20 and buried the dead and sent the wounded to Fort Bayard.

One thing that attracted my attention more than anything else was the suffering of Private John W. Williams, Troop H, who was shot through the kneecap and had to ride all that night from the battle-ground to Brookman's Mill. Poor fellow!

I buried all my dead, and then marched for Fort Cumming, where we arrived about sunset and reported to General Edward Hatch, then commanding the regiment and also the district of New Mexico, giving him all the details pertaining to the engagement.

General Hatch asked me about how many men I could mount the next morning, the 21st. I informed him about how many. He ordered me to have my troop in readiness by daylight and report to Lieutenant Demmick, then commanding Troop L, and follow that Indian trail.

My troop was ready as ordered, and marched. We followed those Indians to the line of Old Mexico, but were unable to overtake them. Such was my last engagement with hostile Indians."

The formula that Negroes cannot command, with the further assertion that colored soldiers will neither follow nor obey officers of their own race, we have now taken out of the heads of its upholders, and away from its secure setting of type on the printed page, and applied it to the facts. Negro soldiers have shown their ability to command by commanding, not always with shoulder-straps, to be sure, but nevertheless commanding. With wearying succession, instance after instance, where Negroes have exercised all manner of military command and always creditably, have extended for us a recital to the border of monotony, and made formidable test of our patience. In France and the West Indies, in Central and South America, Negroes have commanded armies, in one instance fighting under Napoleon, at other times to free themselves from slavery and their countries from the yoke of oppression. In our own country, from the days of the Revolution, when fourteen American officers declared in a memorial to the Congress, that a "Negro man called Salem Poor, of Colonel Frye's regiment, Captain Ames' company, in the late battle at Charlestown, behaved like an *experienced officer*, as well as an excellent soldier;"* from the first war of the nation down to its last, Negro soldiers have been evincing their capacity to command. In the Civil War, where thousands of colored soldiers fought for the Union, their ability to command has been evidenced in a hundred ways, on scouts and expeditions, in camp and in battle; on two notable occasions, Negro officers gallantly fought their commands side by side with white officers, and added lustre to the military glory of the nation. Upon the re-organization of the Regular Army

*MS. Archives of Massachusetts, Vol. 180, p. 241, quoted in Williams's Negro Troops in the Rebellion, p. 13.

at the close of the war the theatre shifted to our Western frontier, where the Negro soldier continued to display his ability to command. Finally, in the Spanish War, just closed, the Negro soldier made the nation again bear witness not alone to his undaunted bravery, but also to his conspicuous capacity to command. Out of this abundant and conclusive array of incontestable facts, frankly, is there anything left to the arbitrary formula that Negroes cannot command, but a string of ipse dixits hung on a very old, but still decidedly robust prejudice? There is no escape from the conclusion that as a matter of fact, with opportunity, Negroes differ in no wise from other men in capacity to exercise military command.

Undoubtedly substantial progress has been made respecting colored officers since 1863, when colored soldiers were first admitted in considerable numbers into the army of the Union. At the period of the Civil War colored officers for colored soldiers was little more than thought of; the sole instance comprised the short-lived colored officers of the three regiments of Louisiana Native Guards, and the sporadic appointments made near the close of the war, when the fighting was over.

More than three hundred colored officers served in the volunteer army in the war with Spain. Two Northern States, Illinois and Kansas, and one Southern State, North Carolina, put each in the field as part of its quota a regiment of colored troops officered throughout by colored men. Ohio and Indiana contributed each a separate battalion of colored soldiers entirely under colored officers.

In 1863 a regiment of colored troops with colored officers was practically impossible. In 1898 a regiment of colored volunteers without some colored officers was almost equally impossible. In 1863 a regiment of colored soldiers commanded by colored officers would have been a violation of the sentiment of the period and an outrage upon popular feelings, the appearance of which in almost any Northern city would hardly fail to provoke an angry and resentful mob. At that period, even black recruits in uniforms were frequently assaulted in the streets of Northern cities. We have seen already how Sergeant Rivers, of the First South Carolina Volunteers, had to beat off a mob on Broadway in New York city. In 1898 regiments and battalions of colored troops, with colored colonels and majors in command, came out of States where the most stringent black laws were formerly in force, and were greeted with applause as they passed on their way to their camps or to embark for Cuba.

In Baltimore, in 1863, the appearance of a Negro in the uniform of an army surgeon started a riot, and the irate mob was not appeased until it had stripped the patriotic colored doctor of his shoulder-straps. In 1898, when the Sixth Regiment of Massachusetts Volunteers passed through the same city, the colored officers of Company L of that regiment were welcomed with the same courtesies as their white colleagues—courtesies extended as a memorial of the fateful progress of the regiment through the city of Baltimore in 1861. One State which went to war in 1861 to keep the Negro a slave, put in the field a regiment of colored soldiers, officered by colored men from the colonel down. To this extent has prejudice been made to yield either to political necessity, or a generous change in sentiment. Thus were found States both North and South willing to give the Negro the full military recognition to which he is entitled.

With this wider recognition of colored officers the general

government has not kept pace. In the four regiments of colored volunteers recruited by the general government for service in the war with Spain, only the lieutenants were colored. Through the extreme conservatism of the War Department, in these regiments no colored officers, no matter how meritorious, could be appointed or advanced to the grade of captain. Such was the announced policy of the department, and it was strictly carried out. The commissioning of this large number of colored men even to lieutenancies was, without doubt, a distinct step in advance; it was an entering wedge. But it was also an advance singularly inadequate and embarrassing. In one of these colored volunteer, commonly called "immune" regiments, of the twelve captains, but five had previous military training, while of the twenty-four colored lieutenants, eighteen had previous military experience, and three of the remaining six were promoted from the ranks, so that at the time of their appointment twenty-one lieutenants had previous military training. Of the five captains with previous military experience, one, years ago, had been a lieutenant in the Regular Army; another was promoted from Post Quartermaster-Sergeant; a third at one time had been First Sergeant of Artillery; the remaining two had more or less experience in the militia. Of the eighteen lieutenants with previous military experience, twelve had served in the Regular Army; eight of these, not one with a service less than fifteen years, were promoted directly from the ranks of the regulars for efficiency and gallantry. At the time of their promotion two were Sergeants, five First Sergeants and one a Post Quartermaster-Sergeant. The four others from the Regular Army had served five years each. Of the six remaining Lieutenants with previous military experience, four had received military training in high schools, three of whom were subsequently officers in the militia; fifth graduated from a state college with a military department; the sixth had been for years an officer in the militia. With this advantage at the start, it is no extravagance to say that the colored officers practically made the companies. To them was due the greater part of the credit for whatever efficiency the companies showed. Moreover, these colored officers were not behind in intelligence. Among them were four graduates of universities and colleges, two lawyers, two teachers, one journalist, five graduates of high schools and academies, and the men from the Regular Army, as their previous non-commissioned rank indicates, were of good average intelligence. There is no reason to believe that this one of the four colored volunteer regiments was in any degree exceptional.

These are the officers for whom the War Department had erected their arbitrary bar at captaincy, and declared that no show of efficiency could secure for them the titular rank which they more than once actually exercised. For they were repeatedly in command of their companies through sickness or absence of their captains. They served as officers without the incentive which comes from hope of promotion. They were forced to see the credit of their labors go to others, and to share more than once in discredit for which they were not responsible. They were, and in this lay their chief embarrassment, without the security and protection which higher rank would have accorded them. In case of trial by court-martial, captains and other higher officers filled the court to the exclusion of almost all others. These were white men. It is gratifying to record that the War Department recognized this special injustice to colored officers, and in the two regiments of colored volunteers recruited for service in the Philippines all the line-officers are colored men, the field officers being white, and appointed from the Regular Army in pursuance of a general policy. Thus far has the general government advanced in recognition of the military capacity of the Negro. In the swing of the pendulum the nation is now at the place where the hardy General Butler was thirty-seven years ago, when he organized the three regiments of Louisiana Native Guards with all line-officers colored.

The way in which modern armies are organized and perfected leaves little necessity for an equipment of exceptional personal gifts in order to exercise ordinary military command. The whole thing is subordinate, and the field for personal initiative is contracted to the minimum. In our own army the President is Commander-in-Chief, and the command descends through a multitude of subordinate grades down to the lowest commissioned officer in the service. We have "Articles of War" and "Regulations," and the entire discipline and government of the army is committed to writing. There is no chance to enshroud in mystery the ability to command. For ordinary military command, with intelligence the chief requisite, little is required beyond courage, firmness and good judgment. These qualities are in no respect natural barriers for colored men.

This last story of the Negro soldier's efficiency and gallantry, told in the pages of this book, teaches its own very simple conclusion. The Cuban campaign has forced the nation to recognize the completion of the Negro's evolution as a soldier in the Army of the United States. The colored American soldier, by his own prowess, has won an acknowledged place by the side of the best trained fighters with arms. In the fullness of his manhood he has no rejoicing in the patronizing paean, "the colored troops fought nobly," nor does he glow at all when told of his "faithfulness" and "devotion" to his white officers, qualities accentuated to the point where they might well fit an affectionate dog. He lays claim to no prerogative other than that of a plain citizen of the Republic, trained to the profession of arms. The measure of his demand—and it is the demand of ten millions of his fellow-citizens allied to him by race—is that the full manhood privileges of a soldier be accorded him. On his record in arms, not excluding his manifest capacity to command, the colored soldier, speaking for the entire body of colored citizens in this country, only demands that the door of the nation's military training school be freely open to the capable of his race, and the avenue of promotion from the ranks be accessible to his tried efficiency; that no hindrance prevent competent colored men from taking their places as officers as well as soldiers in the nation's permanent military establishment.

APPENDIX.

The correspondence following shows the progress of the negotiations for the surrender of the city of Santiago and the Spanish Army, from the morning of July 3d until the final convention was signed on the sixteenth of the same month. This surrender virtually closed the war, but did not restore the contending nations to a status of peace. Twenty-three thousand Spanish soldiers had laid down their arms and had been transformed from enemies to friends. On the tenth of August following, a protocol was submitted by the President of the United States, which was accepted by the Spanish cabinet on the eleventh, and on the twelfth the President announced the cessation of hostilities, thus closing a war which had lasted one hundred and ten days. On the tenth of December a Treaty of Peace between the United States and Spain was signed at Paris, which was subsequently ratified by both nations, and diplomatic relations fully restored. The war, though short, had been costly. One hundred and fifty million dollars had been spent in its prosecution, and there were left on our hands the unsolved problem of Cuba and the Philippines, which promised much future trouble.

Within a month from the signing of the convention, the Army of Invasion, known as the Fifth Army Corps, was on its homeward voyage, and by the latter part of August the whole command was well out of Cuba. Well did the soldiers themselves, as well as their friends, realize, as the former returned from that campaign of a hundred days, that war in the tropics was neither a pastime nor a practice march. The campaign had tested the powers of endurance of the men to its utmost limit. The horrors of war were brought directly to the face of the people, as the ten thousand invalids dragged their debilitated forms from the transports to their detention camps, or to the hospitals, some too helpless to walk, and many to die soon after greeting their native shores. Those who had been so enthusiastic for the war were now quiet, and were eagerly laying the blame for the sorrow and suffering before them upon the shoulders of those who had conducted the war. Few stopped to think that a good part of this woe might be justly charged to those who had constantly resisted the establishment of an adequate standing army, and who, with inconsistent vehemence, had urged the nation into a war, regardless of its military equipment. The emaciated veterans arriving at Montauk were spoken of as the evidences of "military incompetency;" they were also evidence of that narrow statesmanship which ignores the constant suggestions of military experience.

Headquarters United States Forces,
Near San Juan River, July 3, 1898—8.30 A. M.

To the Commanding General of the Spanish Forces, Santiago de Cuba.

Sir:—I shall be obliged, unless you surrender, to shell Santiago de Cuba. Please inform the citizens of foreign countries, and all the women and children, that they should leave the city before 10 o'clock to-morrow morning.

Very respectfully, your obedient servant,
WILLIAM R. SHAFTER,
Major-Genera,l U. S. V.

Reply.

Santiago de Cuba, July 3, 1898.

His Excellency the General Commanding Forces of United States, near San Juan River.

Sir:—I have the honor to reply to your communication of to-day, written at 8.30 A. M., and received at 1 P. M., demanding the surrender of this city, or, in contrary case, announcing to me that you will bombard this city, and that I advise the foreigners, women and children that they must leave the city before 10 o'clock to-morrow morning. It is my duty to say to you that this city will not surrender, and that I will inform the foreign consuls and inhabitants of the contents of your message.

Very respectfully,

JOSE TORAL,
Commander-in-Chief, Fourth Corps.

Headquarters Fifth Army Corps,
Camp near San Juan River, Cuba, July 4, 1898.

The Commanding General, Spanish Forces, Santiago de Cuba, Cuba.

Sir:—I was officially informed last night that Admiral Cervera is now a captive on board the U. S. S. Gloucester, and is unharmed. He was then in the harbor of Siboney. I regret also to have to announce to you the death of General Vara del Rey at El Caney, who, with two of his sons, was killed in the battle of July 1st. His body will be buried this morning with military honors. His brother, Lieutenant-Colonel Vara del Rey, is wounded and a prisoner in my hands, together with the following officers: Captain Don Antonio Vara del Rey, Captain Isidor Arias, Captain Antonio Mansas, and Captain Manuel Romero, who, though severely wounded, will all probably survive.

I also have to announce to you that the Spanish fleet, with the exception of one vessel, was destroyed, and this one is being so vigorously pursued that it will be impossible for it to escape. General Pando is opposed by forces sufficient to hold him in check.

In view of the above, I would suggest that, to save needless effusion of blood and the distress of many people, you may reconsider your determination of yesterday. Your men have certainly shown the gallantry which was expected of them.

I am, sir, with great respect,

Your obedient servant,

WILLIAM R. SHAFTER,

Major-General, Commanding United States Forces.

Headquarters Fifth Army Corps,

Camp near San Juan River, Cuba, July 4, 1898.

To the Commanding General, Spanish Forces, Santiago de Cuba, Cuba.

Sir:—The fortune of war has thrown into my hands quite a number of officers and private soldiers, whom I am now holding as prisoners of war, and I have the honor to propose to you that a cartel of exchange be arranged to-day, by which the prisoners taken by the forces of Spain from on board the Merrimac, and any officers and men of the army who may have fallen into our hands within the past few days, may be returned to their respective governments on the terms usual in such cases, of rank for rank. Trusting that this will meet with your favorable consideration, I remain,

Very respectfully, your obedient servant,

WILLIAM R. SHAFTER,

Major-General, Commanding United States Forces.

Headquarters Fifth Army Corps,

Camp near San Juan River, Cuba, July 4, 1898.

To the Commanding Officer, Spanish Forces, Santiago.

Sir:—It will give me great pleasure to return to the city of Santiago at an early hour to-morrow morning all the wounded Spanish officers now at El Caney who are able to be carried and who will give their parole not to serve against the United States until regularly exchanged. I make this proposition, as I am not so situated as to give these officers the care and attention that they can receive at the hands of their military associates and from their own surgeons; though I shall, of course, give them every kind treatment that it is possible to do under such adverse circumstances. Trusting that this will meet with your approbatoin, and that you will permit me to return to you these persons, I am,

Your very obedient servant,

WILLIAM R. SHAFTER,

Major-General, Commanding United States Forces.

Army of the Island of Cuba,
Fifth Corps, General Staff.

To His Excellency the Commander-in-Chief of the American Forces.

Excellency:—I have the honor to reply to the three communications of your Excellency, dated to-day, and I am very grateful for the news you give in regard to the generals, chiefs, officers and troops that are your prisoners, and of the good care that you give to the wounded in your possession. With respect to the wounded, I have no objection to receiving in this place those that your Excellency may willingly deliver me, but I am not authorized by the General-in-Chief to make any exchange, as he has reserved to himself that authority. Yet I have given him notice of the proposition of your Excellency.

It is useless for me to tell you how grateful I am for the interest that your Excellency has shown for the prisoners and corpse of General Vara del Rey, giving you many thanks for the chivalrous treatment.

The same reasons that I explained to you yesterday, I have to give again to-day—that this place will not be surrendered.

I am, yours with great respect and consideration,

(Signed) JOSE TORAL.

In Santiago de Cuba, July 4, 1898.

Headquarters Fifth Army Corps,
Camp near San Juan River, Cuba, July 6, 1898.

To the Commander-in-Chief, Spanish Forces, Santiago de Cuba.

Sir:—In view of the events of the 3d instant, I have the honor to lay before your Excellency certain propositions to which, I trust, your Excellency, will give the consideration which, in my judgment, they deserve.

I inclose a bulletin of the engagement of Sunday morning which resulted in the complete destruction of Admiral Cervera's fleet, the loss of six hundred of his officers and men, and the capture of the remainder. The Admiral, General Paredes and all others who escaped alive are now prisoners on board the Harvard and St. Louis, and the latter ship, in which are the Admiral, General Paredes and the surviving captains (all except the captain of the Almirante Oquendo, who was slain) has already sailed for the United States. If desired by you, this may be confirmed by your Excellency sending an officer under a flag of truce to Admiral Sampson, and he can arrange to visit the Harvard, which will not sail until to-morrow, and obtain the details from Spanish officers and men on board that ship.

Our fleet is now perfectly free to act, and I have the honor to state that unless a surrender be arranged by noon of the 9th instant, a bombardment will be begun and continued by the heavy guns of our ships. The city is within easy range of these guns, the eight-inch being capable of firing 9,500 yards, the thirteen-inch, of course, much farther. The ships can so lie that with a range of 8,000 yards they can reach the centre of the city.

I make this suggestion of a surrender purely in a humanitarian spirit. I do not wish to cause the slaughter of any more men, either of your Excellency's forces or my own, the final result under circumstances so disadvantageous to your Excellency being a foregone conclusion.

As your Excellency may wish to make reference of so momentous a question to your Excellency's home government, it is for this purpose that I have placed the time of the resumption of hostilities sufficiently far in the future to allow a reply being received.

I beg an early answer from your Excellency.

I have the honor to be,

Very respectfully, your obedient servant,

W. R. SHAFTER,

Major-General, Commanding.

Headquarters Fifth Army Corps,

Camp near Santiago, July 9, 1898.

Hon. Secretary of War, Washington, D. C.

I forwarded General Toral's proposition to evacuate the town this morning without consulting any one. Since then I have seen the general officers commanding divisions, who agree with me in that it should be accepted.

1st. It releases at once the harbor.

2d. It permits the return of thousands of women, children and old men, who have left the town, fearing bombardment, and are now suffering fearfully where they are, though I am doing my best to supply them with food.

3d. It saves the great destruction of property which a bombardment would entail, most of which belongs to Cubans and foreign residents.

4th. It at once releases this command while it is in good health for operations elsewhere. There are now three cases of yellow fever at Siboney in a Michigan regiment, and if it gets started, no one knows where it will stop.

We lose by this, simply some prisoners we do not want and the arms they carry. I believe many of them will desert and return to our lines. I was told by a sentinel who deserted last night that two hundred men wanted to come, but were afraid our men would fire upon them.

W. R. SHAFTER,
Major-General, United States Volunteers.

Reply.

Washington, D. C., July 9, 1898.

Major-General Shafter, Playa, Cuba.

In reply to your telegram recommending terms of evacuation as proposed by the Spanish commander, after careful consideration by the President and Secretary of War, I am directed to say that you have repeatedly been advised that you would not be expected to make an assault upon the enemy at Santiago until you were prepared to do the work thoroughly. When you are ready this will be done. Your telegram of this morning said your position was impregnable and that you believed the enemy would yet surrender unconditionally. You have also assured us that you could force their surrender by cutting off their supplies. Under these circumstances, your message recommending that Spanish troops be permitted to evacuate and proceed without molestation to Holguin is a great surprise and is not approved. The responsibility for the destruction and distress to the inhabitants rests entirely with the Spanish commander. The Secretary of War orders that when you are strong enough to destroy the enemy and take Santiago, you do it If you have not force enough, it will be despatched to you at the earliest moment possible. Reinforcements are on the way of which you have already been apprised. In the meantime, nothing is lost by holding the position you now have, and which you regard as impregnable.

Acknowledge receipt. By order of the Secretary of War.

(Signed) H. C. CORBIN, Adjutant-General.

Headquarters United States Forces,
Camp near San Juan River, Cuba, July 11, 1898.

To His Excellency, the Commander-in-Chief of the Spanish Forces, Santiago de Cuba.

Sir:—With the largely increased forces which have come to me, and the fact that I have your line of retreat securely within my hands, the time seems fitting that I should again demand of your Excellency the surrender of Santiago and your Excellency's army. I am authorized to state that should your Excellency so desire, the Government of the United States will transport your entire command to Spain. I have the honor to be,

Very respectfully, your obedient servant,

WILLIAM R. SHAFTER,
Major-General, Commanding.

Reply.

Army of the Island of Cuba, Fourth Corps,
July 11, 1898.

To His Excellency, the Commander-in-Chief of the Forces of the United States, in the Camp of the San Juan.

Esteemed Sir:—I have the honor to advise your Eminence that your communication of this date is received, and in reply desire to confirm that which I said in my former communication, and also to advise you that I have communicated your proposition to the General-in-Chief. Reiterating my sentiments, I am,

Very respectfully, your obedient servant,

(Signed) JOSE TORAL,
Commander-in-Chief of the Fourth Corps and Military Governor of Santiago.

Headquarters Fifth Army Corps,
Camp near Santiago de Cuba, July 12, 1898.

To His Excellency, Commander-in-Chief of Spanish Forces, Santiago de Cuba.

Sir:—I have the honor to inform your Excellency that I have already ordered a suspension of hostilities, and I will repeat that order, granting in this manner a reasonable time within which you may receive an answer to the message sent to the Government of Spain, which time will end to-morrow at 12 o'clock noon.

I think it my duty to inform your Excellency that during this armistice I will not move any of my troops that occupy the advanced line, but the forces that arrived to-day and which are debarking at Siboney require moving to this camp.

I wish that your Excellency would honor me with a personal interview to-morrow morning at 9 o'clock. I will come accompanied by the Commanding General of the American army, and by an interpreter, which will permit you to be accompanied by two or three persons of your staff who speak English. Hoping for a favorable answer, I have the honor to be,

Very respectfully, your obedient servant,

WILLIAM R. SHAFTER,

Major-General, Commanding.

Army of the Island of Cuma, Fourth Corps,

Santiago de cuba, July 12, 1898—9 P. M.

To His Excellency, the General of the American Troops.

Esteemed Sir:—I have the honor to answer your favor of this date, inform your Excellency that in deference to your desires I will be much honored by a conference with his Excellency, the Commanding General of your army, and your Excellency, to-morrow morning at the hour you have seen fit to appoint.

Very respectfully, your obedient servant,

(Signed) JOSE TORAL,

Commander-in-Chief of the Fourth Army Corps.

Preliminary agreement for the capitulation of the Spanish forces which constitute the division of Santiago de Cuba, occupying the territory herein set forth, said capitulation authorized by the Commander-in-Chief of the Island of Cuba, agreed to by General Toral and awaiting the approbation of the Government at Madrid, and subject to the following conditions:

Submitted by the undersigned Commissioners—

Brigadier-General Don Frederick Escario, Lieutenant-Colonel of Staff Don Ventura Fontan and Mr. Robert Mason, of the city of Santiago de Cuba, representing General Toral, commanding spanish forces, to Major-General Joseph Wheeler, U. S. V., Major-General H. W. Lawton, U. S. V., and First Lieutenant J. D. Miley, Second Artillery, A. D. C., representing General Shafter, commanding American forces, for the capitulation of the Spanish forces comprised in that portion of the Island of Cuba east of a line passing through Aserradero, Dos Palmas, Palma Soriano, Cauto Abajo, Escondida, Tanamo and Aguilera, said territory being known as the Eastern District of Santiago, commanded by General Jose Toral.

1. That pending arrangements for capitulation all hostilities between American and Spanish forces in this district shall absolutely and unequivocally cease.

2. That this capitulation includes all the forces and war material in said territory.

3. That after the signing of the final capitulation the United States agrees, with as little delay as possible, to transport all the Spanish troops in said district to the Kingdom of Spain, the troops, as near as possible, to embark at the port nearest the garrison they now occupy.

4. That the officers of the Spanish Army be permitted to retain their side arms, and both officers and enlisted men their personal property.

5. That after final capitulation the Spanish authorities agree without delay to remove, or assist the American Navy in removing, all mines or other obstructions to navigation now in the harbor of Santiago and its mouth.

6. That after final capitulation the commander of the Spanish forces deliver without delay a complete inventory of all arms and munitions of war of the Spanish forces and a roster of the said forces now in the above-described district, to the commander of the American forces.

7. That the commander of the Spanish forces, in leaving said district, is authorized to carry with him all military archives and records pertaining to the Spanish Army now in said district.

8. That all of that portion of the Spanish forces known as Volunteers, Movilizados and Guerillas, who wish to remain in the Island of Cuba are permitted to do so under parole not to take up arms against the United States during the continuance of the war between Spain and the United States, delivering up their arms.

9. That the Spanish forces will march out of Santiago de Cuba with honors of war, depositing their arms thereafter at a point mutually agreed upon, to await their disposition by the United States Government, it being understood that the United States Commissioners will recommend that the Spanish soldier return to Spain with the arms he so bravely defended.

Entered into this fifteenth day of July, eighteen hundred and ninety-eight, by the undersigned Commissioners, acting under instructions from their respecting commanding generals.

(Signed)

JOSEPH WHEELER,
Major-General U. S. Vols.;
H. W. LAWTON,
Major-General U. S. Vols.;
J. D. MILEY,
1st Lieut. 2d Art., A. D. C. to General Shafter.

FREDERICO ESCARIO,
VENTURA FONTAN,
ROBERT MASON.

Army of the Island of Cuba, Fourth Corps,
Santiago de Cuba, July 12, 1898—9 P. M.

To His Excellency, the General-in-Chief of the American Forces,

Esteemed Sir:—As I am now authorized by my Government to capitulate, I have the honor to so advise you, requesting you to designate the hour and place where my representatives should appear, to concur with those of your Excellency to edit the articles of capitulation on the basis of what has been agreed upon to this date.

In due time I wish to manifest to your Excellency my desire to know the resolution of the United States Government respecting the return of the arms, so as to note it in the capitulation; also for their great courtesy and gentlemanly deportment I wish to thank your Grace's representatives, and in return for their generous and noble efforts for the Spanish soldiers, I hope your Government will allow them to return to the Peninsula with the arms that the American army do them the honor to acknowledge as having dutifully defended.

Reiterating my former sentiments, I remain,

Very respectfully, your obedient servant,

JOSE TORAL,

Commander-in-Chief of the Fourth Army Corps.

At Neutral Camp, near Santiago, Under a Flag of Truce,
July 14, 1898.

Recognizing the chivalry, courage and gallantry of Generals Linares and Toral, and of the soldiers of Spain who were engaged in the battles recently fought in the vicinity of Santiago de Cuba, as displayed in said battles, we, the undersigned officers of the United States army, who had the honor to be engaged in said battle, and are now a duly organized commission, treating with a like commission of officers of the Spanish army, for the capitulation of Santiago de Cuba, unanimously join in earnestly soliciting the proper authority to accord to these brave and chivalrous soldiers the privilege of returning to their country bearing the arms they have so bravely defended.

JOSEPH WHEELER,
Major-General, U. S. Vols.
H. W. LAWTON,
Major-General, U. S. Vols.
First Lient., 2d Art., A. D. C.
J. D. MILEY,

Army of the Island of Cuba, Fourth Corps,
Santiago de Cuba, July 16, 1898.

To His Excellency, the Commander-in-Chief of the Forces of the United States.

Esteemed Sir:—At half-past 11 I received your communication of this date, and I am sorry to advise you that it is impossible for my representatives to come to the appointed place at midday, as you wish, as I must meet them and give them their instructions.

If agreeable to you, will you defer the visit until 4 P. M. to-day or until 7 to-morrow morning, and in the meanwhile the obstacles to the entrance of the Red Cross will be removed from the harbor.

I beg your Honor will make clear what force you wish me o retire from the railroad, as, if it is that in Aguadores, I would authorize the repair of the bridge at once by your engineers; and if it is that on the heights to the left of your lines, I beg you will specify with more precision.

I have ordered those in charge of the aqueduct to proceed at once to repair it with the means at their command.

Awaiting your reply, I remain,

Very respectfully, your obedient servant,

JOSE TORAL,

Commander-in-Chief of the Fourth Army Corps.

Headquarters Fifth Army Corps,

Camp, July 16, 1898.

To His Excellency, General Jose Toral, Commanding Spanish Forces in Eastern Cuba.

Sir:—I have the honor to acknowledge the receipt of your Excellency's letter of this date, notifying me that the Government at Madrid approves your action, and requesting that I designate officers to arrange for and receive the surrender of the forces of your Excellency. This I do, nominating Major-General Wheeler, Major-General Lawton, and my aide, Lieutenant Miley. I have to request that your Excellency at once withdraw your troops from along the railway to Aguadores, and from the bluff in rear of my left; also that you at once direct the removal of the obstructions at the entrance to the harbor or assist the navy in doing so, as it is of the utmost importance that I at once get vessels loaded with food into the harbor.

The repair of the railroad will, I am told, require a week's time. I shall, as I have said to your Excellency, urge my Government that the gallant men your Excellency has so ably commanded have returned to Spain with them the arms they have wielded. With great respect, I remain,

Your obedient servant and friend,

WILLIAM R. SHAFTER,

General, Commanding.

Terms of the Military Convention for the capitulation of the Spanish forces occupying the territory which constitutes the Division of Santiago de Cuba and described as follows: All that portion of the Island of Cuba east of a line passing through Aserradero, Dos Palmas, Cauto Abajo, Escondida, Tanamo and Aguilara, said troops being in command of General Jose Toral; agreed upon by the undersigned Commissioners: Brigadier-General Don Federico Escario, Lieutenant-Colonel of Staff Don Ventura Fontan, and as Interpreter, Mr. Robert Mason, of the city of Santiago de Cuba, appointed by General Toral, commanding the Spanish forces, on behalf of the Kingdom of Spain, and Major-General Joseph Wheeler, U. S. V., Major-General H. W. Lawton, U. S. V., and First Lieutenant J. D. Miley, Second Artillery, A. D. C., appointed by General Shafter, commanding the American forces on behalf of the United States:

1. That all hostilities between the American and Spanish forces in this district absolutely and unequivocally cease.

2. That this capitulation includes all the forces and war material in said territory.

3. That the United States agrees, with as little delay as possible, to transport all the Spanish troops in said district to the Kingdom of Spain, the troops being embarked, as far as possible at the port nearest the garrison they now occupy.

4. That the officers of the Spanish Arm be permitted to retain their side arms, and both officers and private soldiers their personal property.

5. That the Spanish authorities agree to remove, or assist the American Navy in removing, all mines or other obstructions to navigation now in the harbor of Santiago and its mouth.

6. That the commander of the Spanish forces deliver without delay a complete inventory of all arms and munitions of war of the Spanish forces in above described district to the commander of the American forces; also a roster of said forces now in said district.

7. That the commander of the Spanish forces, in leaving said district, is authorized to carry with him all military

archives and records pertaining to the Spanish Army now in said district.

8. That all that portion of the Spanish forces known as Volunteers, Movilizados and Guerillas, who wish to remain in the Island of Cuba, are permitted to do so upon the condition of delivering up their arms and taking a parole not to bear arms against he United States during the continuance of the present war between Spain and the United States.

9. That the Spanish forces will march out of Santiago de Cuba with the honors of war, depositing their arms thereafter at a point mutually agreed upon, to await their disposition by the United States Government, it being understood that the United States Commissioners will recommend that the Spanish soldier return to Spain with the arms he so bravely defended.

10. That the provisions of the foregoing instrument become operative immediately upon its being signed.

Entered into this sixteenth day of July, eighteen hundred and ninety-eight, by the undersigned Commissioners, acting under instructions from their respective commanding generals and with the approbation of their respective governments.

(Signed)

JOSEPH WHEELER,
Major-General U. S. Vols.;
H. W. LAWTON,
Major-General U. S. Vols.;
J. D. MILEY,
1st Lieut. 2d Art., A. D. C. to General Shafter.

FREDERICO ESCARIO,
VENTURA FONTAN,
ROBERT MASON.

The following dispatch, sent by General Linares, will show how desperate were the straits into which he had been driven and how earnestly he desired to be granted authority to avoid further fighting by the surrender of his forces at Santiago:

Santiago de Cuba, July 12, 1898.

The General-in-Chief to the Secretary of War.

Although prostrated in bed from weakness and pain, my mind is troubled by the situation of our suffering troops, and therefore I think it my duty to address myself to you, Mr. Secretary, and describe the true situation.

The enemy's forces very near city; ours extended fourteen kilometres (14,000 yards). Our troops exhausted and sickly in an alarming proportion. Cannot be brought to the hospital—needing them in trenches. Cattle without fodder or hay. Fearful storm of rain, which has been pouring continuously for past twenty-four hours. Soldiers without permanent shelter. Their only food rice, and not much of that. They have no way of changing or drying their clothing. Our losses were very heavy—many chiefs and officers among the dead, wounded and sick Their absence deprives the forces of their leaders in this very critical moment. Under these conditions it is impossible to open a breach on the enemy, because it would take a third of our men who cannot go out, and whom the enemy would decimate. The result would be a terrible disaster, without obtaining, as you desire, the salvation of eleven maimed battalions. To make a sortie protected by the division at Holguin, it is necessary to attack the enemy's lines simultaneously, and the forces of Holguin cannot come here except after many long days' marching. Impossible for them to transport rations. Unfortunately, the situation is desperate. The surrender is imminent, otherwise we will only gain time to prolong our agony. The sacrifice would be sterile, and the men understand this. With his lines so near us, the enemy will annihilate us without exposing his own, as he did yesterday, bombarding by land elevations without our being able to discover their batteries, and by sea the fleet has a perfect knowledge of the place, and bombards with a mathematical accuracy. Santiago is no Gerona, a walled city, part of the mother country, and defended inch by inch by her own people without distinction—old men and women who helped with their lives, moved by the holy idea of freedom, and with the hope of help, which they received. Here I am alone. All the people have fled, even those holding public offices, almost without exception. Only the priests remain, and they wish to leave the city to-day, headed by their archbishop. These defenders do not start now a campaign full of enthusiasm and energy, but for three years they have been fighting the climate, privations and fatigue, and now they have to confront this critical situation when they have no enthusiasm or physical strength. They have no ideals, because they defend the property of people who have deserted them and those who are the allies of the American forces.

The honor of arms has its limit, and I appeal to the judgment of the Government and of the entire nation whether these patient trops have not repeatedly saved it since May 18th—date of first bombardment. If it is necessary that I sacrifice them for reasons unknown to me, or if it is necessary for some one to take responsibility for the issue foreseen and announced by me in several telegrams, I willingly offer myself as a sacrifice to my country, and I will take charge of the command for the act of surrender, as my modest reputation is of small value when the reputation of the nation is at stake.

(Signed) LINARES.

Thus surrendered to our forces about 23,500 Spanish troops, of whom about 11,000 had been in the garrison of Santiago, the others having been stationed in garrisons outside of the city, but belonging to the Division of Santiago. With them were also surrendered 100 cannon, 18 machine guns and over 25,000 rifles. The troops were all sent back to Spain in vessels of their own nation and flying their own flag. We had lost in battles with them before the surrender 23 officers killed and 237 men; and 100 officers and 1,332 men wounded.

CLASSICS *in* BLACK STUDIES SERIES

Anna E. Dickinson, *What Answer?*

Frederick Douglass,
My Bondage and My Freedom

W. E. B. Du Bois, *Darkwater*

W. E. B. Du Bois, *The Negro*

T. G. Steward, *Buffalo Soldiers: The Colored Regulars in the United States Army*

Booker T. Washington and Others,
The Negro Problem

Ida B. Wells-Barnett, *On Lynchings*